GENETICS

THE MACMILLAN COMPANY
NEW YORK · BOSTON · CHICAGO
DALLAS · SAN FRANCISCO

MACMILLAN & CO., Limited
LONDON · BOMBAY · CALCUTTA
MELBOURNE

THE MACMILLAN CO. OF CANADA, Ltd.
TORONTO

GENETICS

AN INTRODUCTION TO THE STUDY
OF HEREDITY

BY

HERBERT EUGENE WALTER

ASSOCIATE PROFESSOR OF BIOLOGY
BROWN UNIVERSITY

WITH 72 FIGURES AND DIAGRAMS

New York
THE MACMILLAN COMPANY
1916

Norwood Press
J. S. Cushing Co. — Berwick & Smith Co.
Norwood, Mass., U.S.A.

THIS VOLUME

IS AFFECTIONATELY DEDICATED

TO

MY MOTHER

PREFACE

THE following pages had their origin in a course
of lectures upon Heredity, given at Brown University during the winter of 1911-1912, which were
amplified and repeated in part the following summer at Cold Spring Harbor, Long Island, before the
biological summer school of the Brooklyn Institute
of Arts and Sciences.

An attempt has been made to summarize for the
intelligent, but uninitiated, reader some of the more
recent phases of the questions of heredity which
are at present agitating the biological world. It is
hoped that this summary will not only be of interest
to the general reader, but that it will also be of service in college courses dealing with evolution and
heredity.

The subject of heredity concerns every one, but
many of those who wish to become better informed
regarding it are either too busily engaged or lack the
opportunity to study the matter out for themselves.
The recent literature in this field is already very
large, with every indication that much more is about
to follow, which is a further discouragement to non-
technical readers.

It may not be a thankless task, therefore, out of
the jargon of many tongues to raise a single voice

which shall attempt to tell the tale of heredity.
There may be a certain advantage in having as
spokesman one who is not at present immersed in the
arduous technical investigations that are making
the tale worth telling.   The difficulties in understanding this complicated subject may possibly be
realized better by one who is himself still struggling
with them, than by the seasoned expert who has
long since forgotten that such difficulties exist.

Among others I am particularly indebted to Dr.
C. B. Davenport for many helpful suggestions, to
my colleague, Professor A. D. Mead, for reading the
manuscript critically, to Dr. S. I. Kornhauser who
gave valuable aid in connection with the chapter
on the Determination of Sex, and to my wife for
assistance in final preparation for the press.

I wish to thank Professor H. S. Jennings and Dr.
H. H. Goddard, who have given generous permission
to copy certain diagrams, as well as The Outlook
Company and The Macmillan Company for the use
of figures 24 and 66, respectively.

The fact that all the suggestions which were at
various times offered by my kindly critics have
not been incorporated in the text, absolves them
from responsibility for whatever remains.

H. E. W.

Providence, R. I.,
    September, 1912.

CONTENTS

ix

CONTENTS

GENETICS

GENETICS

CHAPTER I

INTRODUCTION

1. The Triangle of Life

WITHIN a generation the center of biological interest has gradually been swinging from the origin of species to the origin of the individual. The nineteenth century was Darwin's century. His monumental work "On the Origin of Species by Means of Natural Selection," which appeared in 1859, not only dominated the biological sciences but also influenced profoundly many other realms of thought, particularly those of philosophy and theology.

Now, at the beginning of the twentieth century, a particular emphasis is being laid upon the study of heredity. The interpretation of investigations along this line of research has been made possible through the cumulative discoveries of many things that were not known in Darwin's day. Trained students have been patiently and persistently bending over improved microscopes, untangling the mysteries of the cell, while an increasing host of investigators, inspired by the Austrian monk Mendel, have been industriously devoting their energies to breeding animals and plants with an insight denied to breeders of preceding centuries.

The study of the origin of the individual, which has grown out of the more general consideration of the origin of species, forms the subject-matter of heredity, or, to use the more definitive word of Bateson, of *genetics*.

It is not with the individual as a whole that

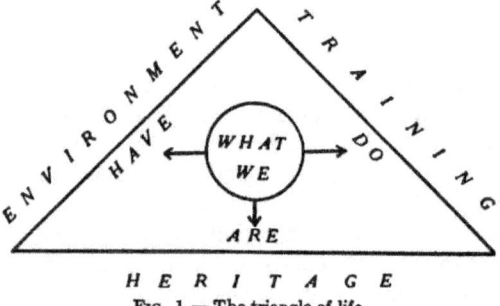

FIG. 1.—The triangle of life.

genetics is chiefly concerned, but rather with *characteristics* of the individual.

Three factors determine the characteristics of an individual, namely, *environment*, *training*, and *heritage* as expressed diagrammatically in Figure 1. It may indeed be said that an individual is the result of the interaction of these three factors since he may be modified by changing any one of them. Although no one factor can possibly be omitted, the student of genetics places the emphasis upon heritage as the factor of greatest importance. Heritage, or "blood," expresses the innate equipment of the individual. It is what he actually *is* even before birth. It is his nature. It is what determines whether he shall be a beast or a man. Consequently in the diagram (Fig. 1), the triangle of life is represented as resting solidly upon the side marked "heritage" for its foundation.

Environment and training, although indispensable, are both factors which are subsequent and secondary. Environment is what the individual *has*, for example, housing, food, friends and enemies, surrounding aids which may help him and obstacles which he must overcome. It is the particular world into which he comes, the measure of opportunity given to his particular heritage.

Training, or education, on the other hand, represents what the individual *does* with his heritage and environment. Lacking a suitable environment a good heritage may come to naught like good seed sown upon stony ground, but it is nevertheless true that the best environment cannot make up for defective heritage or develop wheat from tares.

The absence of sufficient training or exercise even when the environment is suitable and the endowment of inheritance is ample will result in an individual who falls short of his possibilities, while no amount of education can develop a man out of the heritage of a beast. Consequently the biologist holds that, although what an individual *has* and *does* is unquestionably of great importance, particularly to the individual himself, what he *is*, is far more important

in the long run.   Improved environment and education may better the generation already born.   Improved blood will better every generation to come.

What, then, is this "blood" or heritage? Exactly what is meant by heredity?

2. A Definition of Heredity

Professor Castle, in his recent book on "Heredity
in Relation to Evolution and Animal Breeding," has
defined heredity as "organic resemblance based on
descent."   The son resembles his father because he
is a "chip off the old block."   It would be still
nearer the truth to say that the son resembles his
father because they are *both chips from the same block*,
since the actual characters of parents are never transmitted to their offspring in the same way that real
estate or personal property is passed on from one
generation to another.   When the son is said to have
his father's hair and his mother's complexion it
does not mean that paternal baldness and a vanishing maternal complexion are the inevitable consequences.

Biological inheritance is more comparable to the
handing down from father to son of some valuable
patent right or manufacturing plant by means of
which the son, in due course of time, may develop an
independent fortune of his own, resembling in character and extent the parental fortune similarly derived
although not identical with it.

So it comes about that "organic resemblance"

between father and son, as well as that which often appears between nephew and uncle or even more remote relatives, is due not to a direct entail of the characteristics in question, but to the fact that the characteristics are "based on descent" from a common source. In other words, an "hereditary character" of any kind is not an entity or unit which is handed down from generation to generation, but is rather a *method of reaction* of the organism to the constellation of external environmental factors under which the organism lives.

To unravel the golden threads of inheritance which have bound us all together in the past, as well as to learn how to weave upon the loom of the future, not only those old patterns in plants and animals and men which have already proven worth while, but also to create new organic designs of an excellence hitherto impossible or undreamed of, is the inspiring task before the geneticist to-day.

3. THE MAINTENANCE OF LIFE

So far as we know, every living thing on the earth to-day has arisen from some preceding form of life.

How the first spark of life began will probably always be a matter of pure speculation. Whether the beginnings of what is called life came through space from other worlds on meteoric wings, as Lord Kelvin has suggested; whether it was spontaneously generated on the spot out of lifeless components; or whether life itself was the original condition of

matter, and the one thing that must be explained is
not the origin of life, but of the non-living, no one
can say. Leaving aside the first speculation as un-
tenable and the third as irrational, since it jars so
sadly with what astronomers tell us of the probable
evolution of worlds, the theory of spontaneous generation seems to be the last resort to which to turn.

In prescientific days this idea of spontaneous
generation presented no great difficulties to our
imaginative and credulous ancestors. John Milton,
with the assurance of an eye-witness, thus described
the inorganic origin of a lion : —

> " The grassy clods now calved ; now half appears
>   The tawny lion, pawing to get free
>   His hinder parts — then springs as broke from bonds,
>   And rampant shakes his brindled mane."
>   (" Paradise Lost," Book VII, line 543.)

Ovid also in his "Metamorphoses," not to mention a
more familiar instance, easily succeeded in creating
mankind from the humble stones tossed by the
juggling hands of Deucalion and Pyrrha.

Although under former conditions on the earth
it might have been possible for life to have originated
spontaneously, and although it may yet be possible
to produce life from inorganic materials in the laboratory or elsewhere, the exhaustive work of Pasteur,
Tyndall and others effectually demonstrated a generation ago that to-day living matter always arises from
preceding living matter and this conclusion is generally accepted as an axiom in genetics.

There are various methods of producing *more* life, given a nest-egg of living substance with which to start. Any organism, whether plant or animal, is continually transforming inorganic and dead material into living tissue. Through the process of repair, for example, an injury to a form as highly developed even as man is frequently made good, if it is not too extensive, as in the case of a skin wound.

When the intake of non-living material is in excess of the outgo, growth results, with the consequence that more living substance is built up than existed before. Thus a fragment of a living sponge or a piece of a begonia leaf are each sufficient to restore a duplicate of the original organism.

A process similar to the repair of the begonia leaf is that employed so effectively in the great groups of the one-celled animals and plants, the Protozoa and Protophyta, by means of which their numbers are maintained. These one-celled organisms multiply by fission, that is, by equal division into halves, and each half then grows to the size of the parent organism from which it sprang. When two daughter protozoans are thus formed, they are essentially orphans because they have no parents, alive or dead. The parental substance in such a process, along with the regulating power necessary to reorganization, goes over bodily into the next generation in the formation of the daughter-cells, leaving usually no remains whatever behind. In primitive forms of this description, continuous life is the natural order, and death, when it does occur, is, as Weismann has pointed out, accidental and quite outside the plan of nature.

In these cases it is easy to see the reason for "organic resemblance" between successive generations. Parent and offspring are successive manifestations of the *same thing*, just as the begonia plant, restored from a fragment of a begonia leaf, is simply an extension of the original plant.

Many modifications of the process of multiplication by fission occur, all of them, however, agreeing in the fundamental principle that the progeny resemble the parents because they are pieces of the parents.

Thus the greening apple maintains its individuality although coming from thousands of different trees, because all of these trees through the asexual process of grafting are continuations of the one original Rhode Island greening tree grown by Dr. Solomon Drowne in the town of Foster, nearly a century ago.

Again, certain fresh-water sponges and bryozoans, quite unlike most of their marine relatives, keep a foothold from year to year within their particular shallow fresh-water habitats by isolating well protected fragments of themselves in the form of *gemmules* and *statoblasts*. These structures may drop to the muddy bottom and live in a dormant condition throughout the icy winter when it would not be possible for the entire organism to survive near the surface.

In order to meet the conditions imposed by winter, however, these fragments have become so modified

as temporarily to lose their likeness to the parent
generation, although readily regaining that likeness
when springtime brings the opportunity. The unity
of two succeeding generations, although interrupted
by the temporary interposition of something apparently different in the form of gemmules or statoblasts, is thus essentially maintained. The bryozoan
colonies of two successive seasons in a fresh-water
pond may be regarded as parts of the same identical
colony, since they present an "organic resemblance
based on descent," although the sole representatives
of the parent colony during midwinter may be the
sparks of life locked up within the statoblasts buried
in the mud.

Similarly, the asexual spores of many plants, such
as molds, mosses and ferns, may be regarded as
gemmules reduced to the lowest terms, namely, to
single cells. As in the preceding cases so in this
instance the resemblance of the offspring which may
arise from these spores, to the parents which produced them, is due to the essential material identity
of two generations.

These illustrations of heredity in its simplest manifestations give the key to "organic resemblance"
higher up in the scale. Sexual reproduction is no
less plainly the direct continuation of life though in
this instance *two* sporelike fragments out of one
generation contribute to form the new individual of
the next generation instead of *one* fragment. In all
cases there is a *material continuity between succeeding
generations*. Offspring become thus an extension of

a single parent or of two parents, while heredity is simply "organic resemblance based on descent."

4. SOMATOPLASM AND GERMPLASM

In forms that reproduce sexually there theoretically occurs a differentiation of the body substance into what Weismann terms *somatoplasm* and *germplasm*.

The somatoplasm includes the body tissues, that is, the bulk of the individual, which is fated in the course of events to complete a life-cycle and die. The germplasm, on the contrary, is the immortal fragment freighted with the power to duplicate the whole organism and which, barring accident, is destined to live on and give rise to new individuals.

The germplasm thus carries potencies for developing both germplasm and somatoplasm, while the somatoplasm, according to this conception, has only the power to reproduce more of its own kind. Moreover, the germplasm is not formed afresh in each generation, neither does it arise anew when the individual reaches sexual maturity, but it is a continuous substance present from the beginning. Although this theory of the continuity of the germplasm has been actually demonstrated in comparatively few instances, all the facts we know concerning the behavior of the germinal substance are consistent with it.

In many of the Protozoa the entire organism is possibly comparable to germplasm, but in all forms of life that are compounded of several cells the germplasm is probably set aside early in the development of the individual, and this remains undifferentiated, or in reserve, like a savings-bank account put by for a rainy day, while the somatoplasm is expended in the immediate demands of the tissues that make up the individual. In one instance at least, that of the nematode worm Ascaris, according to Boveri, this splitting off or isolation of the germplasm occurs as early in the cleavage of the fertilized egg as the sixteen-cell stage, when fifteen of the cells go to form the somatoplasm and the sixteenth is set aside as germplasm.

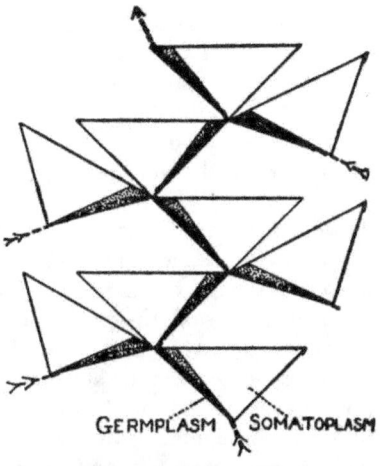

GERMPLASM   SOMATOPLASM

Fig. 2. — Scheme to illustrate *the continuity of the germplasm*. Each triangle represents an individual made up of *germplasm* (dotted) and *somatoplasm* (undotted). The beginning of the life cycle of each individual is represented at the apex of the triangle where germplasm and somatoplasm are both present. As the individual develops each of these component parts increases. In sexual reproduction the germplasms of two individuals unite into a common stream to which the somatoplasm makes no contribution. The continuity of the germplasm is shown by the heavy broken line into which run collateral contributions from successive sexual reproductions.

Thus there results a continuous stream of germplasm, receiving contributions from other germplasmal streams at the time of sexual reproduction, as shown diagrammatically in Figure 2, in which individuals are represented by triangles. From this continuous stream of germplasm there split off at successive intervals complexes of somatoplasm, or "individuals," which go so far on the road of specialization into tissues that the power to be "born again" is lost, and so after a time they die, while the germplasm, held in reserve, lives on.

This is what is meant by saying that a father and son owe their mutual resemblance to the fact that they are chips off the same block rather than by saying that the son is a chip off the paternal block. Both somatoplasms are developments *at different intervals* from the same continuous stream of germplasm instead of one somatoplasm being derived from a preceding one. As a matter of fact the germplasm from which the son arises is modified by the addition of a maternal contribution, so that father and son in reality hold the same relation to each other that half-brothers do.

From the point of view of genetics, then, the real mission of the somatoplasm, which is so marvelously differentiated into all the various forms that we call animals and plants, is simply to serve as a temporary domicile for the immortal germplasm. Thus the parent becomes as it were the "trustee of the germplasm," but not the producer of the offspring.

In the light of these preliminary explanations it is plain that the hopeful point of attack in the science of genetics must inevitably be the germplasm which is the source, or point of departure, in the formation of each new individual, rather than the somatoplasm, which represents the end stages of the hereditary processes.

This has not been the method of the past. The resemblances of the visible father and son have usually been traced instead of the character of their unseen germplasms. By following this old method, investigators have often been misled because the visible or apparent is not always the true index of what lies behind it. A gray and a white rabbit, for example, may produce some offspring that are entirely black just as two white-flowering sweet peas when crossed may sometimes produce purple blossoms. Consequently it is a great fallacy to affirm that in heredity "like produces like," since the opposite is quite often the case.

The new heredity, embodied in the science of genetics, attempts to go deeper than the surface appearance of the somatoplasm. It aims to get at the source or origin of organisms, that is, the *germplasm* which is the only connecting thread between succeeding generations of living forms. It is concerned not so much with somatoplasm, which represents what the germplasm has done in the past, as with the germplasm and what it can do in the future.

CHAPTER II

THE CARRIERS OF THE HERITAGE

1. INTRODUCTION

HEREDITY, as has been shown in the introductory chapter, is essentially a matter of continuity between succeeding generations of living organisms. This continuity may be direct, as when a mother protozoan divides into two daughters, or it may be indirect, as illustrated by the relationship of a father and son, an uncle and nephew, or any other relatives of varying degrees of kinship which, taken singly or collectively, are somatoplasms derived from a common stream of germplasm.

It is the purpose of the present chapter to consider this material continuity between succeeding generations and to discover, if possible, just what are the carriers of the heritage from one generation to another. To this end it will be necessary in the first place to take up what is meant by the "cell theory."

2. THE CELL THEORY

In 1838–1839 the "cell theory" of Schleiden and Schwann, which affirms that all organisms, both plant and animal, are made up of cellular units, had its birth.

14

Robert Hooke, as early as 1665, had described "little boxes or cells distinguished from one another" which he saw in thin slices of cork, and to him is due the rather unfortunate use of the term "cell" which has survived in biological writings to this day. The reason this term is unfortunate is because walls, which are ordinarily the characteristic feature of any cell, such as a prison cell, are usually the least important part of the structure of a living cell, often indeed being entirely absent.

3. A Typical Cell

A typical undifferentiated cell is represented diagrammatically in Figure 3. Near the center of the cell the *nucleus* is shown surrounded by a

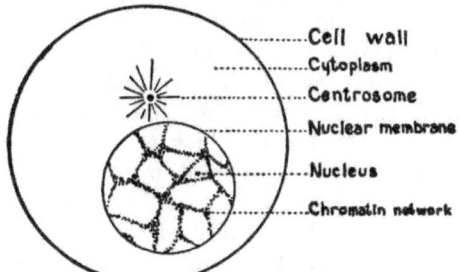

FIG. 3.—Diagram of a typical cell.

nuclear membrane. The nucleus, in common with the enveloping *cytoplasm*, is made up of living substance called *protoplasm* (Hugo von Mohl, 1846), and around the whole there is usually formed a

wall or membrane which serves to separate one cell
from another. Within the protoplasm there may
be a considerable amount of non-living substance
in the form of salts, pigments, oil-drops, water, and
other inclusions of various kinds.

The nucleus is to be regarded as the headquarters
of the whole cell, since changes which the cell undergoes seem to be initiated in it, while cells deprived of
their nuclei cannot long survive.   A single instance
will serve to show the vital part which the nucleus
plays in the life-history of the cell.   In 1883, Gruber
found that after rocking a thin cover-glass back and
forth in a drop of water containing a collection of the
protozoan *Stentor*, which has a long chain-like nucleus,
these tiny animals could thus be cut into fragments,
which would in some instances recover from the
operation and regenerate into complete individuals.
Only those pieces, however, which contained a fragment of the nucleus regenerated into new Stentors,
while pieces of relatively large size which lacked a fragment of nuclear substance very soon disintegrated.

The nucleus, it should be said, is made up of more
than one substance, a fact that is easily demonstrated
by processes of staining, in which certain dyes,
through chemical union, stain a part but not the
whole of the nuclear substance. The part most
easily stained is called *chromatin*, that is "colored
material," and during certain phases of cell life the
chromatin masses together within the nucleus into
visibly definite structures or bodies termed *chromosomes.*

Throughout all the various cells that make up the individuals of any one species these chromosomes appear to be practically constant in number with some exceptions to be mentioned later in connection with sex.   This law of the constant chromosome number for any species was first stated by Boveri in 1900.

The chromosomes of different organisms vary in number from two in the worm Ascaris up to perhaps 1600, according to Haecker ('09), in certain radiolaria. Species which apparently are closely related may differ widely with respect to the number of their chromosomes, while species of unquestionably remote relationship may have an identical number of chromosomes in each of their cells.   The number of chromosomes characteristic for a species, therefore, is in no way an index to the complexity or degree of differentiation of the species.

Besides the nucleus there may often be identified in the cytoplasm of the animal cell a tiny body known as the *centrosome*.   At certain times in the life-cycle of a cell the centrosome becomes the focal point of peculiar radiating lines, which play an important part in the behavior of the cell, particularly during the period of division.

Every cell passes through a cycle of life which may be compared with that common to individuals.   It is born from another cell; passes through a vigorous youth characterized by growth and transformation; attains maturity when the metamorphoses of its earlier life give place to a considerable degree of stability; and finally, after a more or less extended

c

period of normal activity old age ensues, and death completes the cycle. In most instances, however, before this final phase is reached, the cell gives place to daughter-cells through fission, after the manner of most protozoans, and a new cell cycle is begun.

Sometimes the road of differentiation has been traveled so far that it is apparently impossible, as in the case of the complicated brain-cells, to retrace these steps of differentiation and begin again. In such instances the outfit of cells provided in the embryo determines the numerical limit of the cells available throughout life. When this supply is exhausted no more cells appear to replace those which have been worn out.

4. MITOSIS

The ordinary process by which two cells are made out of one is termed *mitosis*. It occurs constantly, and particularly during growth, in all cellular organisms. A series of diagrams, modified from Boveri, illustrating the typical phases of mitosis is given in Figures 4 to 13.

The resting cell (Fig. 4) is characterized by the presence of a nuclear membrane, a single centrosome, and by a chromatin network within the nucleus. In the *beginning of the prophase* (Fig. 5) the centrosome has divided into two parts, while in the *early prophase* (Fig. 6) the two centrosomes have moved farther apart and definite separate chromosomes have formed out of the chromatin network. The *prophase proper* (Fig. 7) is marked by the vanishing of the nuclear

membrane and the more compact form of the chromosomes.   At the *end of the prophase* (Fig. 8) the chromosomes have come to lie at the equator of the cell,

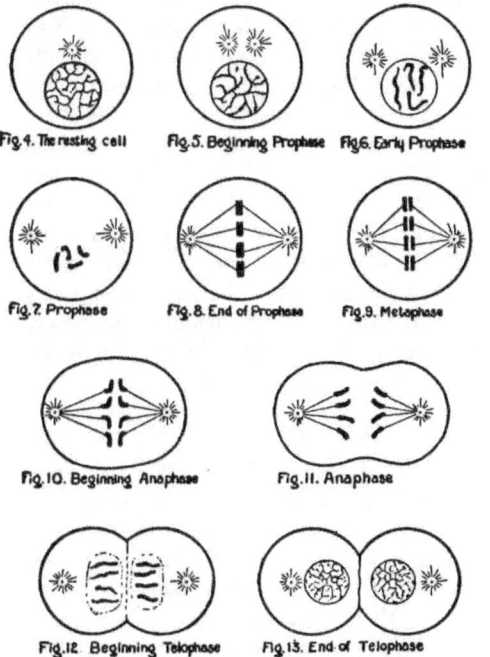

Fig. 4. The resting cell   Fig. 5. Beginning Prophase   Fig. 6. Early Prophase

Fig. 7. Prophase   Fig. 8. End of Prophase   Fig. 9. Metaphase

Fig. 10. Beginning Anaphase   Fig. 11. Anaphase

Fig. 12. Beginning Telophase   Fig. 13. End of Telophase

Figs. 4–13. — Diagrams illustrating mitosis. After Boveri.

being connected by the mantle fibers with the centrosomes, each of which has now come to occupy a
polar position.   In the *metaphase* (Fig. 9) the chromosomes split lengthwise, and at the *beginning of the*

anaphase (Fig. 10) these half chromosomes commence
to separate from each other and to move toward
the poles, while the mantle fibers shorten. During
the *anaphase* (Fig. 11) the cell body lengthens and
begins to divide, while the migration of the half
chromosomes toward the poles is completed. In
the *beginning of the telophase* (Fig. 12) the half
chromosomes grow until they attain full size and
the division of the cell body into two parts becomes
complete. The mantle fibers have disappeared and
the nuclear membrane begins to re-form around the
chromosomes. Finally, at the *end of the telophase*
(Fig. 13) the nuclear membrane becomes complete,
the chromosomes break up into a chromatin network,
and two resting cells take the place of the single one
with which the process began (Fig. 4).

5. AMITOSIS

Amitosis, or the formation of two cells from one
without the machinery of mitosis, is comparatively
rare. It occurs in certain rather isolated instances
among animals and plants, particularly in old cells
late in their life-cycle or in cells that are on the road
to degeneration. When amitosis takes the place
of the more elaborate process of mitosis it is frequently, though not always, a signal of the death-
warrant for that particular cell.

6. SEXUAL REPRODUCTION

The mechanism by means of which two cells unite
to make one in sexual reproduction is quite as com-

plicated as that of mitosis by which one cell is transformed into two.

In sexual reproduction there are two kinds of germ-cells, the egg and the spermatozoan respectively, which take part in producing a new organism. These cells are structurally unlike each other in nearly every particular, but each is a true cell, which von Kölliker made clear as early as 1841, and each has typically the same number of chromosomes in its nucleus, a fact more recently determined by van Beneden in 1883.

The egg-cell is often supplied with one or more envelopes of protective or nutritive function, and it is usually distended with stored up yolk, in consequence of which it is comparatively large and stationary. The result is that whatever locomotion is necessary to bring the two cells together for union devolves upon the sperm-cell. Consequently the sperm-cells are practically nuclei with locomotor tails of cytoplasm, and frequently, in addition, with some structural modification for boring a way into the egg-cell. They are, moreover, much more numerous than the egg-cells, so that although many go astray, never fulfilling their mission, the chances are nevertheless good that some one of them will reach the egg and effect fertilization.

Ordinarily only one sperm enters the egg, but when several succeed in penetrating into the cytoplasm only one proceeds to combine with the egg nucleus, that is, only one sperm nucleus is normally concerned in the essential process of fertilization.

It was formerly thought by the school of "ovists" that in fertilization the essential process is a stimulation of the all important egg by the sperm. The opposing school of "spermists," on the other hand, regarded the egg simply as a nutritive cell the function of which is to harbor the all important sperm. It is now known that both the egg- and the sperm-cell are equally concerned in fertilization, which consists in the union of their respective nuclei within the cytoplasm of the egg.

7. MATURATION

Certain preliminary changes of a preparatory nature, termed *maturation*, regularly precede the union of the nuclei of the two sex-cells in fertilization.

These maturing changes result in reducing the outfit of chromosomes in each sex-cell to one half the original number, a process which is necessary in order to maintain the chromosome count which is characteristic for any particular species and which is known to exist unbroken from generation to generation. If there were no such reduction, then the fertilized egg, formed by the union of egg and sperm nuclei, would contain double the characteristic number of chromosomes, and during the formation of a new individual, the number in all the cells arising by mitosis from such a fertilized egg would likewise be double. When the germ-cells of such individuals unite in fertilization, the original number of chromosomes would be quadrupled, and so on in

geometric progression throughout subsequent generations.   In 1883, too late for Darwin to learn of it,
van Beneden discovered the important fact that the

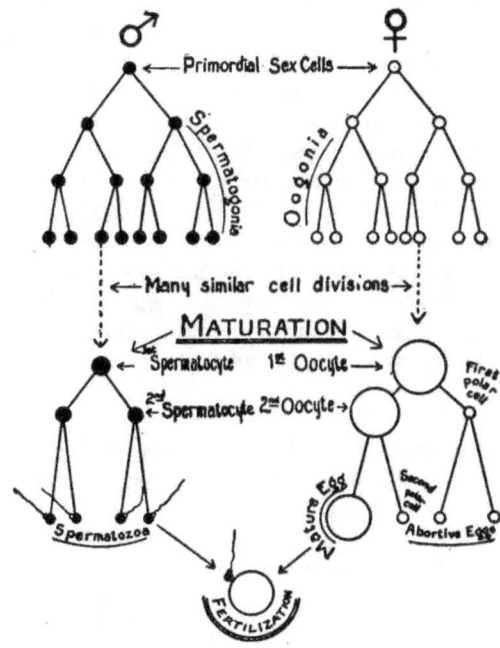

FIG. 14. — Scheme to illustrate maturation of germ-cells.

contain only

l.

If its normal
nete (marry-

ing cell), while the fertilized egg which is formed by the union of two gametes (mature egg- and sperm-cell), and which consequently has the characteristic number of chromosomes, is called a *zygote* (yoked cell).

A diagrammatic representation of the process of maturation is shown in Figure 14. The number of chromosomes (not shown in the diagram) remains constant in each germ-cell respectively until the division of second spermatocytes into spermatids which are subsequently transformed into spermatozoa, and of the second oocytes into mature eggs and second polar cells, when it is reduced to one half the normal number. As spermatozoan and mature egg unite in fertilization, the original number of chromosomes is restored in the fertilized egg (zygote).

8. FERTILIZATION

The stages concerned in a typical case of fertilization, according to Boveri, are illustrated in Figures 15 to 23.

In Figure 15 the "head" and the "middle piece" of the sperm-cell have penetrated into the egg cytoplasm, while in Figure 16 the tail of the sperm-cell has become lost and the middle piece, which furnished the centrosome, has rotated 180° so that it lies between the nucleus, or head, of the sperm-cell and that of the egg-cell. Figure 17 shows an increase in the size of the sperm nucleus and a division of the centrosome into two parts which begin to migrate towards the poles. This process of polar migration of the centrosomes is carried further in Figure 18 as

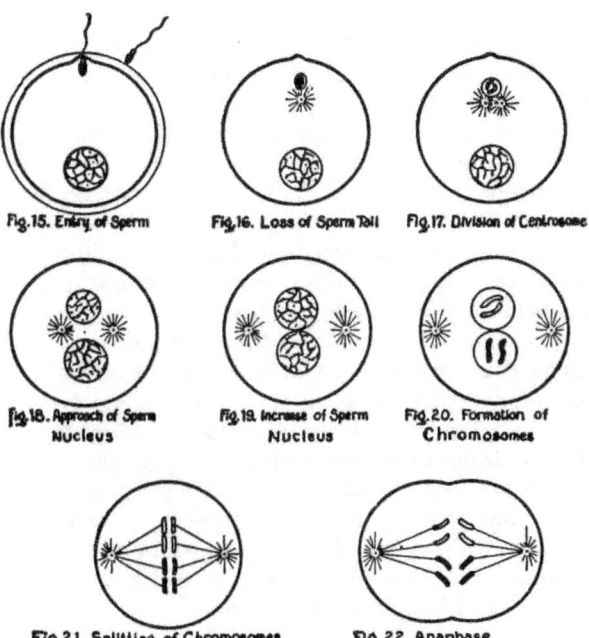

Fig.15. Entry of Sperm   Fig.16. Loss of Sperm Tail   Fig.17. Division of Centrosome

Fig.18. Approach of Sperm   Fig.19. Increase of Sperm   Fig.20. Formation of
Nucleus   Nucleus   Chromosomes

Fig.21. Splitting of Chromosomes   Fig.22. Anaphase

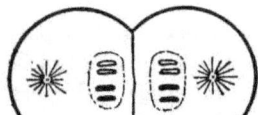

Fig.23. Two-celled Stage

FIGS. 15–23. — Diagrams illustrating fertilization. After Boveri.

well as the increase in the size of the sperm nucleus,
until in Figure 19 the process is complete so that the
centrosomes have assumed a polar position and the
sperm nucleus is equal in size to the egg nucleus
and lies in contact with it. In Figure 20 the chromatin network of the two nuclei has formed into an
equal number of chromosomes which in each case
is half the number characteristic for the species.
Figure 21 shows the complete disappearance of the
nuclear membrane, a process that had already begun
in the preceding figure, and also the arrangement of
the chromosomes, connected with mantle fibers, in the
equatorial plane where the former split longitudinally.
In Figure 22, when the half chromosomes thus formed
pull apart and migrate toward the poles, the segmenta-
tion of the fertilized egg has begun, and there finally
occurs, as shown in Figure 23, the two-celled stage
following fertilization in which each cell contains the
normal number of chromosomes, half of which came
from the egg and half from the sperm.

9. Parthenogenesis

Fertilization is by no means an essential process in
the formation of a new individual, even in those animals which produce both eggs and sperms. Many
animals and plants reproduce parthenogenetically,
that is, the egg-cell may develop without first uniting
with a sperm-cell. In these instances the chromosomes of the egg are not halved during maturation,
and the offspring, therefore, have the same number

of chromosomes as the parent, since they are simply fragments of the parent.

Professor Loeb, by the use of certain chemicals, has succeeded in doing artificially what apparently is never accomplished in nature, namely, making an egg that normally requires fertilization develop parthenogenetically.

10. The Hereditary Bridge

Whatever may ultimately prove to be determiners of the hereditary characters which appear in successive generations, it is obvious that, in any event, such determiners must be located in the zygote, that is, in the fertilized egg. This single cell is the actual bridge of continuity between any parental and filial generation. Moreover, it is the *only* bridge.

In the majority of animals the egg develops entirely outside of and independent of the mother, thus limiting to the egg-cell itself all possible maternal contributions to the offspring. Although there is abundant evidence that half of the filial characteristics come from the male parent, the only actual fragment of the paternal organism given over to the new individual is the single sperm-cell, which unites with the egg in fertilization, and the whole of this even is not usually concerned in the process of fertilization. The entire factor of heritage is packed into the two germ-cells derived from the respective parents and, in all probability, into the nuclei of these germ-cells, since the nuclei are apparently the only portions of these cells that invariably take part in fertilization. To the new individual developing by mitosis from the fertilized egg into an independent organism, the factors of environment and training referred to in Figure 1 are subsequently added.

When it is remembered that the human egg-cell is only about $\frac{1}{125}$th of an inch in diameter, a gigantic *or* *200 micra* size as compared with that of the human sperm-cell, and, furthermore, when one passes in rapid review the marvelous array of characteristics which make up the sum total of what is obviously inherited in man, the wonder grows that so small a bridge can stand such an enormous traffic. A sharp-eyed patrol of this bridge as the strategic focus of heredity is proving to be one of the most effective points of attack in the entire campaign of genetics.

It is not desirable at this time to discuss possible ways in which the determiners of the heritage, whatever they may be, are originally packed into the germ-cells, for this question can be more conveniently considered in a later connection. It is important at present, however, to emphasize the obvious conclusion that determiners of heredity must inevitably be present in the germ-cells in order to account for the fact of "organic resemblance based on descent" between parents and their progeny.

11. The Determiners of Heredity

What are the determiners of hereditary qualities? Do they actually exist in the germ-cells as visible entities, and is there such a thing as a mechanical basis for heredity as the German embryologist Wilhelm His suggested years ago when he wrote: "It is a piece of unscientific mysticism to suppose that heredity will build up an organism without mechanical means"? Can we find these determiners by the aid of microscopes and differential stains, or are they some sort of intangible entities, such as enzymes or hormones or the like, which only the chemist can detect?

Whatever the answer to these questions, it may at least be affirmed that the determiner *represents the adult structure without resembling it.* It is something which controls the unfolding of the developing organism with respect to both quantity and quality, and which also governs the time and rate of appearance of its various characteristics so that certain combinations rather than others shall come about in definite sequence. To use the words of Conklin: "The mechanism of heredity is the mechanism of differentiation."

12. THE CHROMOSOME THEORY

Certain investigators, who seek a morphological basis for heredity, regard the chromosomes as the carriers of the heritage; in other words, as the source of the determiners of ontogeny or the effective factors in the process of differentiation.

A few of the grounds for this theory are briefly indicated below.

First: In spite of the great relative difference in

size between the egg-cell and the sperm-cell, in heredity the two are practically equivalent, as has been repeatedly shown by making reciprocal crosses between the two sexes. The only features that are apparently alike in both the germ-cells are the chromosomes. The inference is, therefore, that they contain the determiners which are the causal factors for the equivalence of adult characters in heredity. The existence of an extra chromosome in probable connection with the matter of sex is, as will be pointed out later, an exception to the exact chromosome equivalence of the two sexes, which only goes to strengthen the supposition that the chromosomes are the carriers of hereditary qualities since extra chromosomes are always associated with the character of sex.

Second: The process of maturation, which always results in halving the chromosome material of the germ-cells as a preliminary step to fertilization, is a series of complicated manœuvers not practised by other cells. During this process no other part of the cells appears to play so consistent and important a rôle as the chromosomes. Provided they act as hereditary carriers, their peculiar behavior during maturation is just what is needed to bring together an entire complement of hereditary determiners out of partial contributions from two parental sources.

Third: Sometimes abnormal fertilization occurs, as in the case when two or more sperm-cells, instead of one, enter the egg cytoplasm and unite with the egg nucleus. This unusual performance has been artificially induced by chemical means in the case of sea-urchins' eggs. The fertilized egg, or zygote, thus formed with an excess of male chromosomes, results in the development of abnormal larvæ. It is thought that a causal connection may exist, therefore, between the additional male chromosomes in the fertilized ovum and the abnormalities of the progeny.

Fourth: The fact that chromosomes may retain their individuality throughout the complicated phases of mitosis, as has been proven in some instances, agrees with the corresponding fact that certain characteristics of the somatoplasm maintain their individuality from generation to generation.

Moreover, certain chromosomes in the fertilized egg have been identified with particular features in the adult developing from that egg. Tennent summarizes his recent work on Echinoderms (1912) by the statement that from a knowledge of the chromosomes in the parental germ-cells, particular characters in the adult hybrids may be predicted, and, conversely, that from the appearance of sexually mature hybrids the character of certain chromosomes in their germ-cells may be predicted.

Again, the correlation of a particular chromosome in the germ-cells with a definite adult character, namely sex, has been repeatedly demonstrated in connection with the so-called "extra chromosome" to which reference has already been made.

Fifth: Finally, excellent evidence of a definite causal connection between certain chromosomes of the germ-cells and particular somatic characters has been furnished by certain critical experiments upon the eggs of sea-urchins. Boveri found that he was able in some instances to shake out the nuclei bodily, chromosomes and all, from the mature eggs of the sea-urchin, *Sphærechinus*, and when there was added in sea water to such enucleated eggs the sperm-cells of an entirely different genus of sea-urchin, namely, *Echinus*, the Echinus sperm-cells entered the *Sphær-echinus* eggs, which had been robbed of their nuclei, and from this peculiar combination larvæ developed which exhibited *only Echinus characters!*

Such cumulative circumstantial evidence as the foregoing has convinced many that in the chromosomes we have visibly before us the carriers of heredity.

Several biologists, however, raise an objecting voice to this theory, protesting against the monopoly of the heritage by the chromosomes. They point out that there always exists an intimate physiological relationship between the nucleus and the cytoplasm, and that it is unreasonable to expect the isolation of one from the other, since the two must always act together as parts of an organic cell unit.

In sexual reproduction, moreover, some small amount at least of spermatic cytoplasm in the form of the so-called "middle piece," which is situated between the head and the tail of the sperm-cell (Fig. 15), may enter the egg about to be fertilized along with the sperm "head" or nucleus, containing the chromosomes. In this way the cytoplasm of the male sperm-cell may not necessarily be entirely excluded from taking part in the formation of the zygote. As a matter of fact, this extra-nuclear part of the sperm-cell sometimes apparently forms the centrosome of the fertilized egg and in consequence may have a hand, as well as the nucleus with the chromosomes, in determining what follows.

13. THE ENZYME THEORY OF HEREDITY

It is not unlikely that the key to this whole problem will be furnished by the biochemists and that the final analysis of the matter of the heritage-carriers will be seen to be chemical rather than morphological in nature.

It has been found that the blood of greyhounds and dachshunds is chemically different, although from a morphological point of view it is apparently identical. The idea of "individual albumen" or "protein specificity" for each animal of a species, to say nothing of the animals of different species, has been advanced as not improbable.

Miescher has shown that an albumen compound having only forty carbon atoms, a number by no means unusual, would make possible a million combinations of atoms or isomers.

The possibilities in this direction seem to be unlimited if we take into consideration those invisible actuators of chemical processes, the *enzymes*, which the chemist brings forward with the prodigality of an astronomer dealing in star-dust, to explain different chemical reactions.

D

Montgomery has suggested that the chromosomes themselves may be masses of enzymes although, according to the chemist, enzymes are not morphological entities, since they seem to be able to flourish and maintain their identity while bringing about chemical reactions in their neighborhood without being visibly demonstrable.

As said before, it is quite likely that in the final analysis heredity will be reduced to a series of chemical reactions dependent upon the manner in which various enzymes initiate, retard, or accelerate successive chemical combinations occurring in the protoplasm. When the same enzymes act upon the same chemical combinations in successive generations, they bring about that "organic resemblance" known as heredity.

E. B. Wilson, whose brilliant work in the entire field of cell activity makes it possible for him to speak with authority, has recently said: "The essential conclusion that is indicated by cytological study of the nuclear substance is, that it is an aggregate of many different chemical components which do not constitute a mere mechanical mixture, but a complex organic system and which undergo perfectly ordered processes of segregation and distribution in the cycle of cell life. That these substances play some definite rôle in determination is not mere assumption, but a conclusion based upon direct cytological experiment and one that finds support in the results of modern chemical research."

14. Conclusion

The supposition that the chromosomes, with certain chemical reservations, are the morphological carriers of the heritage forms an excellent working hypothesis, and this chapter may suitably be closed with a second quotation from Professor Wilson. "In my view studies in this field are at the present time most likely to be advanced by adopting the comparatively simple hypothesis that the nuclear substances are actual factors of reaction by virtue of their specific chemical properties; and I think that it has already helped us to gain a clearer view of some of the most puzzling problems of genetics."

CHAPTER III

VARIATION

1. The Most Invariable Thing in Nature

In the introductory chapter it was shown that "organic resemblance based on descent," by which is meant heredity, is due principally to the fact that offspring are material continuations of their parents and consequently may be expected to be like them. The fact that this is the case in the great majority of instances has given rise to the popular formula, "like produces like," as a rule of heredity.

But this formula by no means always fits the facts. Like often produces something apparently unlike. For instance, two brown-eyed parents may produce a blue-eyed child, although brown-eyed children are more usual from such a parentage. It is a common experience, indeed, for breeders of plants and animals to meet with continual difficulties in getting organisms to "breed true."

On the other hand, it is exactly these variations which so constantly interfere with breeding true that furnish the sole foothold for improvement. If all organisms did breed strictly true, one generation could not stand on the shoulders of the preceding generation, and there would be no evolutionary advance.

36

The most invariable thing in nature is variation.
This fact is at once the hope and the despair of the
breeder who seeks to hold fast to whatever he has
found that is good and at the same time tries to
find something better. When the similarities and
dissimilarities between succeeding generations are
clear, then heredity can be explained. The entire
subject of variation is intimately and inevitably
bound up with any consideration of genetics.

2. THE UNIVERSALITY OF VARIATION

Much of the variation in nature is patent to the
most casual observer, but it requires a trained eye to
see the universal extent of many minor differences.
A flock of sheep may all look alike to a passing stran-
ger, but not to the man who tends them. A dozen
blue violet plants from different localities might
easily be identified by the amateur botanist as belonging to the same species when, to a specialist
on the genus Viola, unmistakable differences would
doubtless be clearly apparent.

The fact that every attempt at an intimate ac-
quaintance with any group of organisms whatsoever
invariably reveals previously unrecognized variations, indicates that variability is much more widespread in nature than is commonly believed.

The key to Japanese art, as pointed out by Dr.
Nitobe, consists in being natural and in faithfully
copying nature. It is for this reason that the Japanese artist makes each object that he produces

unique, because nature herself, whom he strives to follow, never duplicates anything.

The Bertillon system of personal identification is based upon the constancy of minor variations found in each individual. Its importance is shown in Figure 24. The faces of the criminals there pictured would be easily confused by the ordinary observer, but an examination of their thumb prints shows unmistakable differences between these three individuals.

3. Kinds of Variation

A brief enumeration of some of the kinds of variation will reveal their diverse character.

a. With respect to their nature variations may be morphological, physiological, or psychological. Under *morphological* variations are included differences in shape, size, or pattern as well as differences in number and relation of constituent parts.

Differences in activity are of a *physiological* nature. Many animals in captivity are less fertile than when free, while different individuals are well known to vary widely with respect to their susceptibility to disease. Nägeli, for example, reports the presence of tubercles in 97 per cent of the cases in five hundred autopsies, although a majority of the deaths in question was not due to tuberculosis at all, — a fact which indicates a great diversity in the resistance of different individuals to the tubercle bacillus.

Psychological variations in man, such as those which determine the disposition or mental traits of individuals, are apparent to every one.

FIG. 24. —The constancy of minor variations. The thumb prints of these three criminals are characteristically different although their faces would easily confuse the ordinary observer. From The Outlook of Feb. 24, 1912.

b. *With respect to their duplication* variations may be single or multiple. A legless lamb [1] is an example of a *single* variation or "sport." Four-leaved clovers, on the contrary, are *multiple* for the reason that this variation, although not common, nevertheless occurs frequently.

c. *With respect to their utility* variations may be useful, indifferent, or harmful to the organism possessing them. *Useful* variations are of the kind emphasized by Darwin as being effectively made use of in natural selection. *Indifferent* variations, on the other hand, are those which apparently do not play an important part in the welfare of their possessor, as, for example, the color of the eyes or of the hair. Finally, the degree of degeneration in certain organs may be cited as an illustration of *harmful* variations. The amount of closure of the opening from the intestine into the vermiform appendix in man is an example of a harmful variation, since the larger the opening, the greater is the liability to appendicitis.

d. *With respect to their direction in evolution* variations may be either *definite* (orthogenetic) or *indefinite* (fortuitous).

Paleontology furnishes numerous instances of the former category, such as the series of variations from a pentadactyl ancestor, all apparently tending in one direction, which have culminated in the one-toed horse. The fact that the paleontologist deals historically with a completed phylogenetic series in which the side lines lack prominence, while the suc-

---

[1] "A Peculiar Legless Lamb." Stockard. Biol. Bull. xiii, p. 288.

cessful line stands out with distinctness, makes it easy for him to view successive variations as orthogenetic, that is, as definitely directed in one course either through intrinsic (Nägeli) or extrinsic (Eimer) causes.

Fortuitous or chance variations in all possible directions furnish the repertory of opportunity, according to Darwin, from which natural selection picks out those best adapted to survive in the struggle for existence.

e. *With respect to their source*, variations may be somatic or germinal. *Somatic*, or body variations, arise as modifications due to environmental factors. They are individual differences which may be quite transitory in nature, while *germinal* variations may arise without regard to the environment, are deep-seated, and of racial rather than of individual significance.

f. *With respect to their normality* variations may fall within expected extremes and thus be considered *normal*, or they may be outside of reasonable expectations and consequently be reckoned as *abnormal*, as in the case of a two-headed calf.

g. *With respect to the degree of their continuity* variations may form a continuous series, grading into each other by intermediate steps, or they may be discontinuous in character. An example of *continuous* variation is the height of any hundred men one might chance to meet, which would probably represent all intermediate grades from the highest among the hundred to the lowest.

The number of segments in the abdomen of a

shrimp, on the other hand, which may, for instance, be either eight or nine but cannot be halfway between, illustrates what is meant by *discontinuous* variation. The widespread occurrence of this later category of variations has been pointed out by Bateson in his encyclopedic volume "On Materials for the Study of Variation."

h. With respect to their character variations may be quantitative or qualitative. A six-rayed starfish represents a *quantitative* variation from the normal number of five rays, whereas a red variety of a flower may differ chemically from a blue variety, or a bitter fruit may differ from a sweet fruit in a *qualitative* way dependent upon the chemical constitution of the fruit in question.

i. With respect to their relation to an average standard variations may have a *fluctuating* distribution around an arithmetical mean, as when some of the offspring have more and some less of the parental character, or the variations in the progeny may all center about a new average quite distinct from the parental standard and consequently come under the head of *mutations*.

j. Finally, and most important in the present connection, *with respect to heritability*, variations may possess the power to reappear in subsequent generations, or they may lack that power. It is this aspect of variability which bears most directly upon genetics.

Other possible categories might be mentioned, but a sufficient number have been cited to show the great diversity of variations in general.

4. Methods of Studying Variations

Roughly stated, there are three ways of studying variations: *first*, Darwin's method of observation and the description of more or less isolated cases; *second*, Galton's biometric method of statistical inquiry; and *third*, Mendel's experimental method. The second of these methods will be considered in this chapter.

5. Biometry

The new science of biometry, that is, the application of statistical methods to biological facts, has been developed within recent years. Sir Francis Galton, Darwin's distinguished cousin, may be regarded as the pioneer in this field of research, while Karl Pearson and his disciples constitute the modern school of biometricians.

Although mathematical analysis of biological data when not sufficiently ballasted by biological analysis of the same facts may sometimes lead the investigator astray, yet often the only way to formulate certain truths or to analyze data of some kinds is by resort to statistical methods. Biometricians are quite right in insisting that it is frequently necessary to go further than the *fact* of variation, which may be apparent from the inspection of an individual case, and to deal with cumulative evidence as presented through statistical analysis.

In matters of heredity, however, facts as they occur in single cases and definite pedigrees seem to offer a more hopeful line of approach than statistical generalizations. It is better to become acquainted with the real parent than to evolve a hypothetical "mid-parent" mathematically. In this connection it is well always to bear in mind the warning of Johannsen, himself a past master in biometry, when he writes : "*Mit* Mathematik nicht als Mathematik treiben wir unsere Studien."

6. Fluctuating Variation

With respect to any measurable character there are bound to be deviations from an average condition. According to the mathematical laws of chance these deviations sometimes are plus and sometimes minus, and consequently they may be termed *fluctuating variations*.

Pearson gives as a simple illustration of fluctuating variation the number of ribs present in two sets of beech-leaves, as shown below. These sets were taken from two different trees, and each contains twenty-six leaves.

Number of Ribs	13	14	15	16	17	18	19	0	Total
First tree . . .			1	4	7	9	4	1	26
Second tree . .	3	4	9	8	2				26
Total	3	4	10	12	9	9	4	1	

It will at once be seen that, while certain leaves might well belong to either tree, as, for example, those with sixteen ribs, the entire group of leaves from

either tree is unlike that of the other tree. In the
first instance the number of ribs fluctuates around

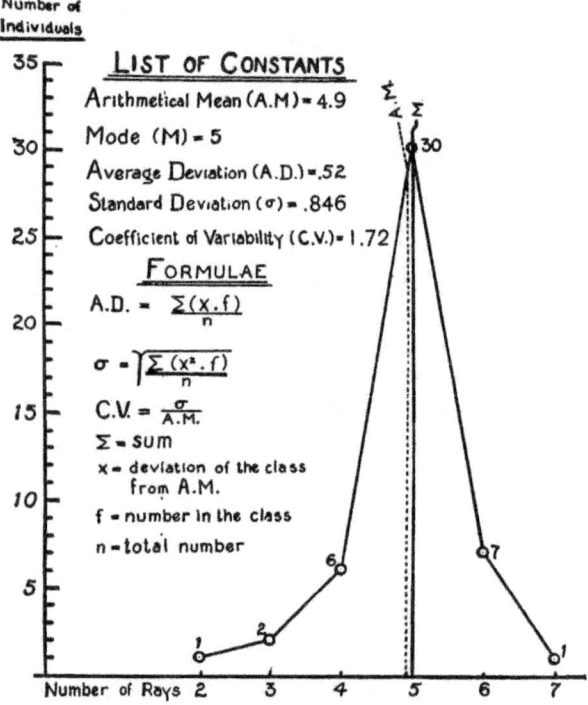

FIG. 25.—The fluctuating variability of starfish rays. From data by
Goldschmidt.

eighteen as the commonest kind; in the second case,
around fifteen. Such a difference could not easily

be detected or expressed by any other method than the statistical one.

Again, in the case of forty-seven starfishes all of which were collected from one locality the variation in the number of rays proved to be, according to Goldschmidt, an amount indicated graphically in Figure 25, where the data are arranged in the form of a so-called frequency polygon or curve.

From such a polygon certain *constants* may be computed which conveniently express in a single number, for purposes of abstract comparison, distinctions that otherwise could be handled only in the most indefinite way.

Thus in this instance the *arithmetical mean*, expressed by the hypothetical number 4.915, a number which of course does not actually occur in nature, is simply the average number of rays in forty-seven starfishes selected at random.

The *mode* which represents the group containing the largest number of individuals of a kind, namely, thirty out of forty-seven, is five in this particular polygon.

The *average deviation*, which is an index of the amount of variation going on among the starfishes in question, is .52. In other words, .52 is the average amount that each individual starfish deviates from the arithmetical mean of 4.915. Although the one seven-rayed starfish which happens to be in the lot varies from the standard of 4.915 to the extent of 2.085 (7 — 4.915) rays, there are thirty five-rayed starfishes which vary only .085 (5 — 4.915) of a ray, and consequently the average of the entire forty-seven amounts to .52 of a ray. In another collection of starfishes where either more seven-rayed or two-rayed specimens might be present, the average deviation would probably be greater.

By computing the average deviation, therefore, and using it as the criterion of variation, a comparison of the variability of organisms that have been taken from different localities or subjected to different conditions can be definitely expressed.

A measure of variability more commonly in use by biometricians, since for mathematical reasons it is more accurate, is the *standard deviation*. This is the square root of the sum of all the deviations squared and their frequencies divided by n, according to the formula

$$\sigma = \sqrt{\frac{\Sigma\,(x^2 \cdot f)}{n}},$$

in which x represents the deviation of each class from the arithmetical mean; f, the number of individuals in each separate class; Σ, the sum of the classes; and n, the total number of individuals.[1]

In the present instance the standard deviation is .846, an arbitrary number that has valuable significance only when brought into comparison with standard deviations similarly derived from other groups of starfishes.

Such a variation polygon as the above expresses the law that the farther any single group is from the

[1] For directions explaining the use of such formulæ see Davenport's "Statistical Methods."

mean of all the groups making up the polygon, the
fewer will be the individuals that represent it.

7. The Interpretation of Variation Polygons

a. Relative Variability

The statistical determination of the relative variability of two lots of organisms with respect to a
certain character may be illustrated by the case of
the oyster-borer snail, *Urosalpinx cinereus*, as seen in
the accompanying table.

ATLANTIC AND PACIFIC SHELLS COMPARED

	LOCALITY	NUMBER OF SHELLS	A. M.	σ	PROBABLE ERROR
Woods Hole	West Shore	1,001	58.928	2.339	±.0352
	Penzance Point	1,002	61.718	2.737	±.0412
	Nobska Point	1,002	61.737	2.152	±.0324
	Nobska Point	1,001	61.944	2.234	±.0337
	Nobska Point	496	66.944	2.366	±.0507
	Barnacle Beach	998	63.932	2.604	±.0393
	Big Wepecket	1,006	57.426	2.052	±.0308
	Mid-Wepecket	500	57.606	2.098	±.0447
	Average for Mass.		61.066	2.335	±.0386
California	Belmont Beds	1,008	59.051	3.023	±.0454
	San Francisco Bay	520	60.892	3.361	±.0703
	Average for Cal.		59.664	3.138	±.0538
	Difference			.803	

The obvious conclusion to be drawn from this
table is that the snails which were unintentionally
carried from the Atlantic coast to California in the

transplantation of oysters show more variation in
their new habitat than they did in the old one with
respect to the particular character measured, namely,
the relative size of the mouth aperture compared with
the height of the entire shell.[1]

b. Bimodal Curves

Sometimes two conspicuous modes make their appearance in a frequency polygon, as Jennings found,

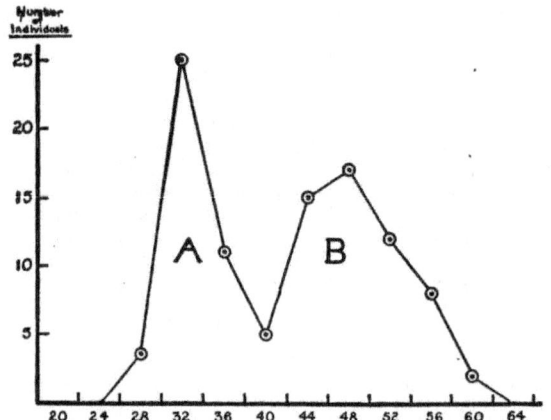

Fig. 26. — The body width of a population of the protozoan *Paramecium*,
showing a polygon with two modes.   A, *Paramecium aurelia.*   B,
Paramecium caudatum.   After Jennings.

for example, in measuring the body width of a population of the protozoan *Paramecium* (Fig. 26).

[1] "Variation in Urosalpinx."   Walter.   Amer. Nat. 1910, Vol. XLIV,
pp. 577–594.

It was subsequently found that the two modes in this polygon were due to the fact that the material in question was a mixture of two closely related species, *Paramecium aurelia* and *Paramecium caudatum*, the individuals of which arranged themselves around their own mean in each instance.

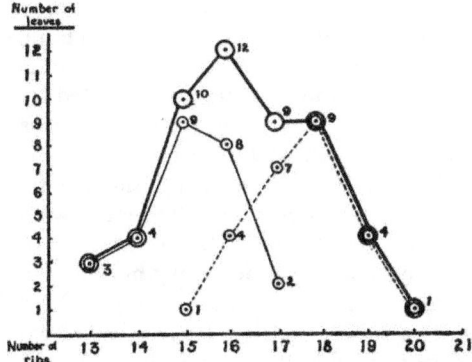

Fig. 27. — The ribs of leaves from two beech trees. When put together they form a polygon which does not reveal its double origin. From data by Pearson.

Although such an explanation does not always turn out to be the right one, the biometrician is led to suspect when a two or more moded polygon appears that he is dealing with a mixture of more than one kind of material, each of which fluctuates around its own average.

Heterogeneous material, it should be noted, does not always give a bimodal curve. For example, if Pearson's two lots of beech leaves mentioned

above are mixed together, they form a regular
series from the inspection of which no one could infer
their double origin. (See the heavy line in Figure 27.)

c. Skew Polygons

The direction in which variations are tending
may sometimes be determined by the statistical
method. As an illustration of this may be cited the
number of ray florets on 1000 white daisies (*Chrysanthemum leucanthemum*), 500 of which were collected at random by the writer from a small patch
in a swampy meadow in northern Vermont, while
the other 500 were selected in the same random
manner upon the same day from a dry hillside pasture hardly more than a stone's throw distant.
Among these two lots of daisies the number of ray
florets varies from twelve to thirty-eight and their
frequency polygons, as shown in Figure 28, form
what are termed "skew polygons," because the mode
in each case lies considerably to one side of the arithmetical mean.

It will be seen that lot *A* from the swampy meadow,
which in spite of the greater fertility of the soil and
the unquestionably greater luxuriance of the plants
themselves, produced heads with fewer florets,
fluctuates around the number 21, while the dry
pasture population *B*, characterized by blossoms
which were in general noticeably smaller, fluctuates
around the number 34.

The habitats of the two lots were so near together, however, that there was probably a considerable intermixture of the two types, as shown by the tendency of each polygon to produce a second mode.

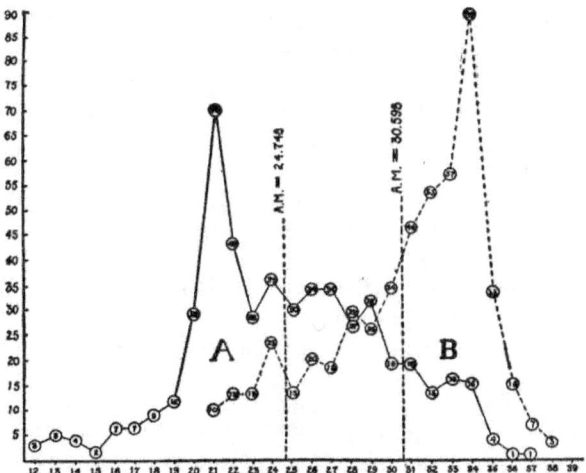

Fig. 28. — Variation in the ray florets of the white daisy (*Chrysanthemum leucanthemum*). *A*, from a swampy meadow. *B*, from a hillside pasture near by. Both the polygons are "skew" because in each case there is an admixture of the other type. The distinction between the two types is due to heredity rather than to environment.

Thus the *A* polygon shows that there is an increasing tendency or variability in the twenty-one floret type toward the thirty-four floret type, due probably in this particular instance to invasion resulting from the proximity of the *B* colony.

8. Graduated and Integral Variations

It is comparatively simple to treat statistically *integral* variations, illustrations of which have been given in the case of beech-leaf ribs, starfish rays, and daisy florets, all of which are characters that can be readily counted. In the same way any measurable character, such as the size of snail shells, may fall into easily limited groups, as, for example, 10 to 11 mm., 11 to 12 mm., 12 to 13 mm., etc. It is somewhat more difficult to classify variations when color or pattern is the character in question, since it then becomes necessary to define certain arbitrary limits for each class of the series within which to group the individual variants.

Tower, in his famous researches on potato-beetles, encountered variations in the pigmentation of the pronotum all the way from entire absence of color to complete pigmentation. By cutting up this continuous series of variations into arbitrary groups of equal extent, however, it was quite possible to arrange the data so that they could be statistically treated just as conveniently as the integral variations mentioned above. Groups or classes of this kind are termed *graduated* variations.

9. The Causes of Variation

With respect to the causes of variation authoritative biologists have taken different points of view.

a. Darwin considered variations as axiomatic. An axiom is self-evident, requiring no explanation.

The absence of variations in organisms rather than the occurrence of variations is, from this point of view, the phenomenon requiring an explanation. Although Darwin himself spent some time in pointing out the universal occurrence of variability, he accepted it as a primary fact and proceeded from it as a starting point without attempting to seek its causes.

b. Lamarck and his followers have regarded the causes of variation either as extrinsic, that is, referable to external factors making up the environment of the organism, or as intrinsic or physiological, that is, based upon the efforts which an organism puts forth to fit into its particular environment successfully. The causes of variation are to be sought ac-

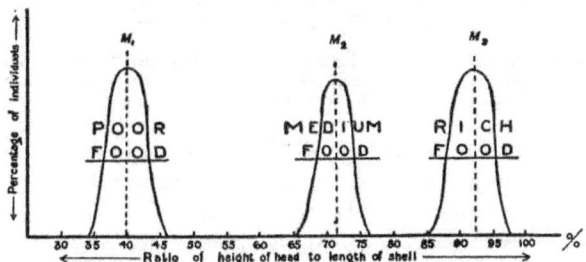

Fig. 29. —Schematic curve of the head height of *Hyalodaphnia* under various conditions of nourishment. Adapted from Woltereck.

cording to the Lamarckian school, in the "environment" and "training" sides of the triangle of life rather than in the "heritage" side (Fig. 1).

For example, Woltereck, by controlling the single

extrinsic factor of food supply, was able to modify
the height of the "head" of the microscopic fresh-
water crustacean, *Hyalodaphnia*, in the remarkable
manner indicated in Figure 29. When poor food

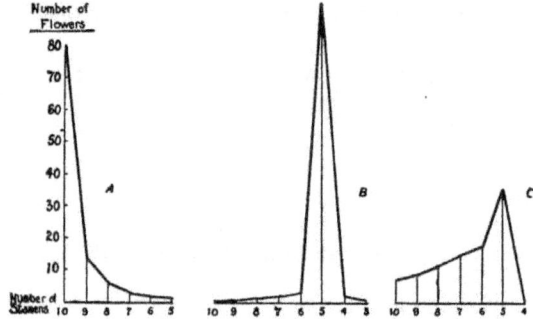

FIG. 30. — Variations in the number of stamens in the flowers of the "live-
for-ever" (*Sedum spectabile*) under various controlled conditions.
For detailed description, see text. After Klebs.

was supplied, the percentage of the head height to
that of the body averaged hardly forty, while with
rich food it was increased to over ninety.

Similarly Klebs succeeded in changing at will the
number of stamens in the common "live-for-ever,"
Sedum spectabile, by manipulating the environment
in which the plants were kept. Some of his results
are shown in Figure 30. Polygon *A* combines the
data for 4260 flowers which were raised in well-fertilized dry soil under bright light; polygon *B* represents 4000 flowers grown in a moist greenhouse
under red light; and polygon *C* includes 4390 flowers

from well-fertilized soil in moist hotbed conditions under a weak light.

c. Weismann, on the contrary, believes that the causes of variation, at least of heritable variations, are intrinsic or inborn in the germplasm. His conception of sexual reproduction is that it is a device for doubling the possible variations in the offspring by the mingling of two strains of germplasm (*amphimixis*). By far the greater number of observations recorded go to substantiate this theory.

Tower found among his potato-beetles, for example, that two strains reared in the same environment showed striking differences in variation, — a fact necessarily due to intrinsic rather than to extrinsic factors. Similar cases may be recalled by any one.

d. Lastly, *Bateson*, whose work "On Materials for the Study of Variation" already cited is a classic, takes the agnostic attitude that it is rather futile to guess at the causes of variation before the facts are well in hand. He consequently discourages such attempts by saying : "Inquiry into the causes of variation is, in my judgment, premature."

In conclusion, the words of Darwin written half a century ago — "Our ignorance of the laws of variation is profound" — may still be appropriately quoted, notwithstanding the fact that in biometry we have at least an excellent analytical method by means of which considerable insight into variation is being gained.

CHAPTER IV

MUTATION

1. The Mutation Theory

AMONG the possible kinds of variation already hinted at are so-called *mutations* which are clearly defined from the fluctuating variations to which reference has just been made.

Darwin was fully aware of the existence of mutations or "sports" and incidentally gave time to their consideration, but the great task which he accomplished in such a masterly manner was to overthrow the widespread and deep-seated belief of his day in a sudden special creation of distinct species. To this end he marshaled evidence in support of the gradual transition of one species into another, emphasizing fluctuations rather than mutations which seemed to him to play a minor rôle in the origin of species.

It remained for the Dutch botanist Hugo de Vries to analyze the character of mutations. There is something distinctly suggestive of Darwin's method in the fact that de Vries worked in silence for twenty years before he gave to the world the "Mutations-theorie" with which his name will forever be connected.

2. Mutation and Fluctuation

A mutation is something qualitatively new that appears abruptly without transitions and which breeds true from the very first. To use the musician's phraseology, it is not a variation elaborated upon an old theme, which would correspond to a fluctuating variation, but it is an entirely new theme. The difference between mutations and fluctuating variations is generally not one of quantity or magnitude, although it sometimes may be so, — since mutations are often much smaller than fluctuations. Mutations are discontinuous in the same sense that chemical combinations, such as carbon monoxide (CO) and carbon dioxide (CO_2), are discontinuous, but the leap from one to the other may be so small that frequently it is difficult to ascertain by inspection alone whether the difference is due to a mutation or a fluctuation. *The test comes in breeding,* for the progeny of a fluctuation will vary around the old average of the parental generation, while the progeny of a mutation will vary around a new average, set by the mutation itself.

When a series of mutations is treated statistically, it does not arrange in frequency polygons as readily as a series of fluctuations do. The latter mass around the average standard according to the laws of chance much in the same way that a hundred shots by a good marksman may center around a bull's-eye. Mutations never act in this way. They find no correspondence even with wild shots at the bull's-eye. They are shots directed at a different target altogether.

To the student of heredity there are two distinctions of prime importance with respect to mutations. First, that they usually appear full-fledged without preparatory stages, and second, that they breed true from the start. Fluctuations, on the contrary, ordinarily "revert" to the parental type in subsequent generations. The great practical importance to the breeder of a knowledge of these heritable mutations is at once apparent.

3. Freaks

A further distinction should be made between mutations and so-called freaks or monstrosities, namely, that the former breed true, while the latter do not. A human physical deformity, such as a club-foot, for example, or a humped back, is not a mutation, because it does not reappear as a heritable character. Variations of this kind are not predetermined in the germplasm, but are usually instances of something that went wrong during the development of the individual somatoplasm.

Thus, among normally "right-handed" snails "left-handed" individuals have occasionally been discovered which, when bred, were found to produce all normal "right-handed" progeny. They are therefore not mutations at all, but freaks or monstrosities due probably to some unusual occurrence early in the cleavage stages of the embryo.

4. KINDS OF MUTATION

De Vries has classified mutations according to their component units into three categories: progressive, regressive, and degressive.

Progressive mutations are signalized by the addition of a new character to the sum of complex characters making up the individual. If rumor may be believed, Anne Boleyn, the second in the interesting series of wives of Henry VIII, was a progressive mutant with respect to at least one character, for she is said to have possessed an extra finger on each hand, as well as the abnormalities of supernumerary mammæ and extra teeth. Evidences that each of these three characters occur as heritable mutations is presented in Davenport's "Heredity in Relation to Eugenics."

Regressive mutations are characterized by the dropping out of something. Thus albinism is caused by the absence of pigment or color. Albinic mutants which breed true are well known, particularly among mammals, such as rats, mice, rabbits, cats, guinea-pigs, and even man himself.

Degressive mutations include cases of the return of a character which was formerly present in the past history of the race, but which has for generations been absent or latent. Castle's four-toed race of guinea-pigs furnishes an example of this class of mutations. In 1906 Professor Castle discovered a newly born guinea-pig in one of his pens with four toes on each hind foot, from which he has successfully established a four-toed race. The hypothetical ancestor of the rodents probably had five toes on each foot, but the normal number in modern guinea-pigs is four on each of the front feet and three on the hind feet. The individual from which Castle has bred a four-toed race exhibited a degressive mutation, tending toward the ancestral type.

5. Species and Varieties

The doctors have always disagreed regarding a definition of species. What determines the exclusive boundaries that shall isolate from their fellows any particular group of animals or plants has long been a mooted question, and still remains so.

The Linnæan concept of a species was that of an exclusive caste of individuals, inflexibly demarked, over whose high barriers no nondescript tramps would dare attempt to climb. When an entomologist of the old Linnæan school encountered an insect which did not conform to the morphological traditions of its fellows, the frequent fate of such a nonconformist was to perish under the boot-heel rather than to find sanctuary in the cabinet of the preserved. Since it was an exception, and a violator of the divine law of the fixity of species, it deserved to be annihilated! Those were hard days both for heretics and for mutations.

The method of the older school of systematists may be described as one which emphasized *differences* and put up barriers that should keep the unlike apart, at the same time allowing only "birds of a feather to flock together." It was a brave and successful attempt to bring order out of chaos by classifying the living world, and it served its purpose well until Darwin's idea of half a century ago, that the origin of all species is from preceding species, put an entirely new face upon the whole matter. Organisms of different species were found to be *related to one another*, and even man could no longer escape acknowledging his poor animal relations. As a consequence, *likenesses rather than differences* thereafter claimed the most attention.

During the reconstruction of phylogenetic trees, which seized the imagination and became the principal business of biologists as soon as the " Origin of Species " was made common property, the crotched sticks in the woodpile of organisms, that had hitherto been largely discarded, were most eagerly sought after. It was just these scraggly sticks, that were neither trunk nor limb-wood but combinations of both, which told the story of continuity and were indispensable in building up a reunited whole.

As the analysis of the living world gradually came to shift from species to individuals, it was shown that individuals may be regarded simply as aggregates of *unit characters* which may combine so variously that it becomes more and more difficult to maintain constant barriers of any kind between the groups of individuals arbitrarily called "species."

The old species of the systematist, upon analysis into their respective unit characters, dissolve into numerous "elementary species" and "varieties" dif-

fering from species perhaps only by the addition
or subtraction of a single character, and thus the

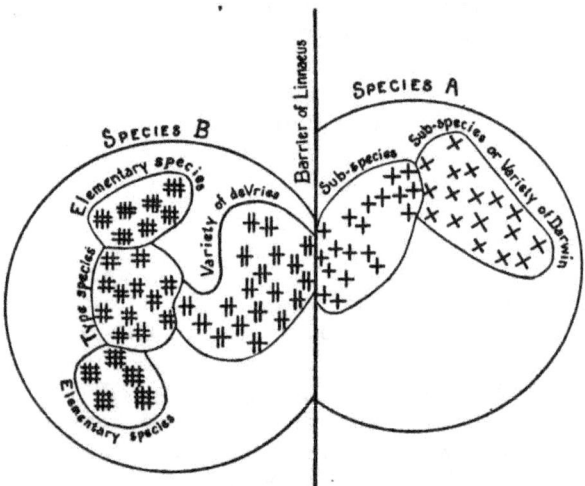

FIG. 31. — Diagram to illustrate various ideas about "species." Under
Species A are represented two groups of individuals which are near
enough alike to be placed within a single species, but which are sufficiently unlike each other to constitute the "sub-species" or "varieties" of Darwin. Under *Species B* are various groups of individuals
distinguished from each other by the addition or loss of one or more
characters. These groups represent the "elementary species" and
"varieties" of de Vries. The "barrier of Linnæus" attempted to
separate species absolutely from each other. Darwin sought to find
loopholes in this barrier. To-day attention is directed rather to the
relation between individuals than to the boundaries between species.

possibilities of analytical classification have become
almost limitless.

An *elementary species*, according to de Vries, is a
progressive mutation differing from the type species

by the addition of at least a single character, while *varieties* are regressive mutations distinguished from the parent type by the loss of at least one character. Both breed true to their respective modifications.

These different concepts of what constitutes a species, illustrated diagrammatically in Figure 31, pave the way for a better understanding of mutations in connection with heredity.

6. Plant Mutations found in Nature

The oldest known authenticated case of a plant mutation is the often cited instance of the "fringed celandine," *Chelidonium laciniatum*, which made its appearance in the garden of the Heidelberg apothecary Sprengel in 1590 among plants of the "greater celandine," *Chelidonium majus*. The fringed celandine bred true at once and is now a widespread and well-known species.

The purple beech has appeared historically as a mutant among ordinary beeches upon at least three occasions in widely separated localities, and it has always given rise to a constant progeny.

The "Shirley poppy," notable for its remarkable range of color, originated from a single plant of the small red poppy, *Papaver rhœas*, which is commonly found in English cornfields.

Instances are known of double flowers among roses, azaleas, stocks, carnations, primroses, petunias, etc., arising from single flowering plants, the seeds of which in turn produce double flowers.

7. Lamarck's Evening Primrose

The most widely known plant mutations are the progeny of Lamarck's evening primrose, *Œnothera lamarckiana*, because it was these plants that led de Vries to formulate his mutation theory.

It is believed by botanists in general that this plant is a native of the southern United States, although it is now, so far as is known, extinct as a wild species in America, and native specimens are included in but few American herbaria.

It was exported to London as a garden plant about 1860, and from thence it spread to the continent, where, escaping from gardens, it became wild in at least one locality near Hilversum, a few miles from Amsterdam. Here, in an abandoned potato field, it fell under the seeing eye of Hugo de Vries in 1885, and now both botanist and primrose are famous.

De Vries found among these escaped plants not only *O. lamarckiana*, but also two other kinds or mutants, *O. brevistylis*, characterized by short-styled flowers, and *O. lævifolia*, which has smooth leaves. These two were entirely new species hitherto unknown at the great botanical clearing-houses of Paris, Leyden, and the Kew Gardens.

Since the seeds of the *Œnothera* are produced by self-fertilized flowers, de Vries felt safe in regarding these plants as mutants rather than hybrids, and he continued to study them with especial care. Transplanting the mutants along with representatives of *O. lamarckiana* to his private gardens in

Amsterdam, where it was possible to maintain them in normal healthy condition, de Vries was able to follow their individual histories with certainty.

He found that, out of 54,343 plants of the species *O. lamarckiana* grown during eight years, there appeared 837 mutants comprising seven different elementary species, all of which, with the exception of *O. scintillans*, bred true. See table.

MUTANTS OF ŒNOTHERA LAMARCKIANA

GENERATION		GIGAS	ALBIDA	OBLONGA	RUBRINERVIS	LAMARCKIANA	NANELLA	LATA	SCINTILLANS	TOTAL
I	1886–7					9				
II	1888–9					15,000	5	5		
III	1890–1				1	10,000	3	3		
IV	1895	1	15	176	8	14,000	60	73	1	
V	1896		25	135	20	8,000	49	142	6	
VI	1897		11	29	3	1,800	9	5	1	
VII	1898			9		3,000	11			
VIII	1899		5	1		1,700	21	1		
		1	56	350	32	53,509	158	229	8	54,343

Some explanatory comment on this table may be of value.

The seeds in each generation were self-fertilized *lamarckiana*.

The mutant *gigas* occurred once, in 1895. From the seeds of this one plant were produced 450 true *gigas* offspring in the first year, and the strain continues to breed true.

Albida was first noted in 1895, but de Vries remem-

F

bered having seen it before and dismissing it as pathological. Because of its poverty in chlorophyll it is a mutant which probably would not maintain itself successfully in nature, but it breeds constant under cultivation.

Oblonga always bred true with the exception of throwing an *albida* in 1895 and a *rubrinervis* in 1899.

Of *rubrinervis* over 2000 invariably bred true, while *nanella* bred true in over 20,000 offspring, with but three exceptions when *oblonga* characters appeared.

Lata, since it produces only female flowers and so cannot be self-fertilized, had constantly to be crossed back with the parent *lamarckiana*, when it produced from 15 to 20 per cent *lata* and 80 to 85 per cent *lamarckiana*.

Finally, *scintillans* which appeared at three separate times proved constant only in its inconstancy because it invariably produces a heterogeneous progeny. The 1895 plant gave 53 per cent *lamarckiana*, 35 per cent *scintillans*, 10 per cent *oblonga*, and 1 per cent *lata*. One of the 1896 plants gave 51 per cent *lamarckiana*, 39 per cent *scintillans*, 8 per cent *oblonga*, 1 per cent *lata*, and 1 per cent *nanella*, while another 1896 plant gave only 8 per cent *lamarckiana*, but 69 per cent *scintillans*, 21 per cent *oblonga*, and 2 per cent of *nanella* and *lata* together.

These seven elementary species are distinguished from each other by features which are unmistakable even to the uninitiated. The old-time systematist would undoubtedly have regarded them as distinct species.

De Vries' experiments and observations have been repeated on a large scale and extended, notably by MacDougal in the New York Botanical Gardens and by Shull at the Carnegie Institution for Experimental Evolution, Cold Spring Harbor, Long Island, and his conclusions have been confirmed in all essential points. The mutability of *O. lamarckiana* is as unmistakable and as diverse in America as it is in Holland.

A parallel case of a plant caught in the act of giving rise to mutations is that of the roadside weed *Lychnis*, reported by Shull, and the phenomenon is probably by no means as unusual as is generally believed. The chief reason why such definite examples of mutation are so infrequently noted and recorded is because the attention of the investigator has generally been directed, not to them, but to gradual fluctuating variations which, according to Darwin's conception, furnish the material for the operation of natural selection. Mutations are doubtless much more common than has been generally supposed, and it is likely that they will receive more attention in the future than they have in the past. De Vries rather pointedly says: "The theory of mutations is a starting-point for direct investigation, while the general belief in slow changes has held back science from such investigations during half a century."

8. Some Mutations among Animals

In 1791 a Massachusetts farmer, by name Seth Wright, found in his flock of sheep a male lamb with long, sagging back and short, bent legs resembling somewhat a German dachshund. With unusual foresight he carefully brought up this strange lamb because it was an animal that could not jump fences. It occurred to this hard-headed Yankee that it would be much easier to get together a flock of short, bow-legged sheep, unable to negotiate anything but a low hurdle, than to labor hard at building high fences. So it came about that this mutating lamb, in the hands of a man who appreciated labor-saving devices, became the ancestor of the Ancon breed of sheep. Later on this breed gave place in public favor to another mutant, the Merino, which produces a superior grade of wool.

Hornless cattle suffer fewer injuries from one another than horned cattle. It has consequently become quite a general practice among farmers to "dehorn" their stock surgically. It is an obvious advantage to have cattle born hornless, and many breeds having this character are now established. In 1889 a mutant among horned stock appeared at Atchison, Kansas, in the form of a hornless Hereford. From this mutant has descended the well-established race of polled Hereford cattle, constituting a bovine aristocracy with registry books and blue blood all their own.

Taillessness in cats, dogs and poultry, as well as hairlessness in cattle, dogs, mice and horses, are further instances of mutations.

Davenport,[1] writing of his experiments with poultry,

[1] Davenport, C. B., 1909. "Inheritance of Characteristics in Domestic Fowl." Carnegie Institution of Washington, Publication No. 121.

says: "During the past four years I have handled
and described over 10,000 poultry of known ancestry.
Of striking new characters I have observed many,
some incompatible with normal existence; others in
no way unfitting the individual for continued life.
In the egg unhatched I have obtained Siamese twins,
pug jaws, and chicks with thigh bones absent. There
have been reared chicks with toes grown together
by a web, without toenails or with two toenails to
a toe; with five, six, seven, or three toes; with one
wing or both lacking; with two pairs of spurs; without oil-gland or tail; with neck devoid of feathers;
with cerebral hernia and a great crest; with feather
shaft recurved, with barbs twisted and dichoto-
mously branched or lacking altogether. Of comb
alone I have a score of forms. All of these characters
have been offered to me without the least effort or
conscious selection on my part, and each appeared in
the first generation as well-developed peculiarities,
and in so far as their inheritance was witnessed, each
refused to blend when mated with a dissimilar form."

Bateson (1894), in his "Materials for the Study of
Variation," gives a detailed list of 886 cases of "discontinuous variations" among animals, many of which
doubtless belong to the category of mutations, although several must be placed in the non-inherit-
able class of "freaks."

9. Possible Explanations of Mutation

It is apparent that the causes of mutations, since
they occur regardless of the environment, are prob-

ably of an intrinsic or germinal nature. Evening primroses display the same mutants whether in Holland or America, in a wild state or under cultivation. Mutations, like poets, are born, not made.

It has been suggested by Standfuss that species may go through the same kind of a life-cycle that in-

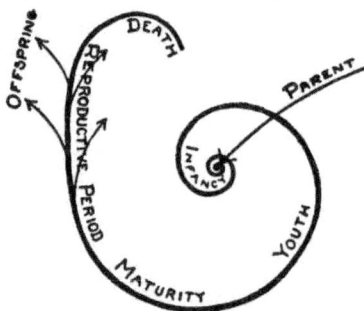

FIG. 32. — Diagram of the relation of reproduction to the life-cycle.

dividuals do, only taking infinitely more time to do it. As shown in Figure 32, they are born of other species and enter the prodigious growth period of infancy and youth, both of which are characterized by much fluctuation.

With maturity they gradually become comparatively stable until the reproductive period is reached, when they throw off their progeny, as on a tangent. They finally pass into the excessively differentiated period of old age, from which there is no recall, although they approach in many features the infantile condition, and end in death or extinction. This cycle is repeatedly illustrated by phylogenetic lines of fossil forms which have long since become extinct.

Beecher has pointed out that, in paleontological times just before they became extinct, species often underwent extreme specialization in the form of fantastic shapes, an excessive number of spines or elaborate sculpturings on the shells as seen among the ammonites, belemnites, and trilobites, or of gigantic size as in the dinosaurs, plesiosaurs, and theromorphs. All of these facts indicate a *species-cycle* in which these abnormal features were the unmistakable signs of old age.

The reproductive period of a species when mutants are being thrown off, as of an individual, may extend over a considerable period of the whole cycle, or it may be confined to a relatively small segment. It is possible that in the evening primrose de Vries may have caught a plant passing through the crucial period of species-reproduction.

Another reason why so few mutations have as yet been seen, is because the majority of organisms are not, during the short span of human observation, in the reproductive part of their cycles. When it is remembered that accurate observation with this object in view has extended over only a brief period, insignificant in comparison with the vast geologic stretches of time concerned in species-building, the marvel is that so much, rather than that so little, has been seen.

A further suggestion in connection with the possible sources of mutation is that mutations may be the results of hybridization, appearing as Mendelian recessives after crossing. As a matter of fact, Sprengel's *Chelidonium laciniatum*, already cited, when crossed with *Chelidonium majus*, behaves according to such an expectation. This phase of the question, however, may be more suitably considered later after what is meant by a "Mendelian recessive" has been made clear.

It is extremely doubtful, however, whether any recombination of parental characters in a hybrid may properly be called a mutation, since no character strictly new is thus produced.

10. A Summary of the Mutation Theory

The main features of the mutation theory of de Vries may be indicated as follows : —

a. New species arise abruptly regardless of environment without transitional forms, and at present they are not known to arise in any other way.

b. New forms arise as unusual deviations from the parent form, which itself remains unchanged, although it may repeatedly give rise to similar deviations.

c. New mutations are, from the first, constant, that is, they produce their like. They do not become gradually established as the result of natural selection.

d. Among mutations there may occur forms characterized by the addition of something new, — *progressive elementary species,* — as well as forms lacking something present in the parental type, — *regressive varieties.*

e. The same mutation may arise simultaneously in many individuals instead of as a single " sport."

f. Mutations do not vary around an arithmetical mean with respect to the parent form, as is the case with fluctuating variations, but they fluctuate around a new average of their own, thus forming a discontinuous series with the parent form.

g. Mutations may occur in all directions, that is, they are not necessarily definite or orthogenetic.

h. Mutations probably appear periodically.

i. Every mutation means a possible doubling of the species.

j. Useless or insignificant fluctuating variations are not necessarily the material from which natural selection must sift out new species.

The bearing of the whole matter of mutation upon heredity lies in the fact that, contrary to Darwin's belief, it is apparently mutations, and not fluctuations, that make up heritable variations. If this supposition proves to be true, mutations furnish the essential material in the study of heredity. Consequently, whatever knowledge we may gain of them has a direct relation to the entire problem of genetics.

CHAPTER V

THE INHERITANCE OF ACQUIRED CHARACTERS

1. SUMMARY OF PRECEDING CHAPTERS

HEREDITARY resemblance is due to the derivation of offspring from the same stock as the parent, and successive generations, therefore, are simply periodic expressions of the same continuous stream of germplasm.

Perfect inheritance, or uniformity of generations, does not exist, since variations always occur in successive generations. It is upon these variations that evolution depends. Without them there would be no change of type and consequently no possibility of evolutionary advance.

Some variations are fluctuating or continuous in character and may be detected and analyzed by statistical methods, while others are mutations, or discontinuous variations, representing qualitative differences which do not lend themselves readily to statistical analysis.

Mutations are more common than was formerly believed, and since they are germinal rather than somatic in character, they play an important rôle in heredity.

74

2. The Bearing of this Chapter upon Genetics

Only those variations which reappear in succeeding generations have to do with heredity. Hence it becomes important to inquire as to what kind of variations actually reappear. Can variations that are not inborn, but which are acquired during the lifetime of the individual, be inherited? Does the experience of the parent become a direct part of the child's heritage, or can the environment of the one enter in any way into the heredity of the other? Can changes wrought in the somatoplasm be so impressed upon the germ-plasm as to change it in such a way that it, in turn, will give rise to similarly modified somatoplasm in the next generation? Can *nurture* as well as *nature* be transmitted?

In answering these questions we are of course concerned solely with *biological inheritance* and not at all with those extra-biological accumulations in the way of arts, literature, tradition, invention, and the like which constitute civilization and which make us the "heirs of the ages." Such benefits are entailed upon us much in the same way as property is "inherited," but they form no part of the personal biological heritage into which we are now inquiring.

3. The Importance of the Question

This inquiry concerning the inheritance of acquired characters, which Professor Brooks has called "the interminable question," is not simply an academic matter. Its solution is of vital importance from several viewpoints. For breeders, who are trying to maintain or improve particular strains of animals or plants; for physicians, who, in fighting disease, are honestly seeking to substitute an ounce of prevention for a pound of cure; for sociologists and philanthropists, who have at heart the permanent bettering of human conditions; for educators, who cherish hopes that their life-work of unfolding the youthful mind may prove cumulative and lasting rather than transitory; for religious workers, who want their faith strengthened that conquests in character-building may outreach the individual and so enrich the race; for parents, who entertain hopes that their own efforts may give their children a better biological start in life, — for all these and many more, it is important to know the answer to the question: Can acquired characters be inherited?

4. An Historical Sketch of Opinion

That the personal accumulations of a lifetime are heritable was generally believed all through the credulous ages. A century ago Lamarck made this idea the corner-stone of his theory of evolution. It was all very simple. The reason evolution occurs in nature is because individual acquirements are being continually added to the onflowing stream of living forms. This cumulation of characters indeed *is* evolution. How else can the present stage of adaptation of organisms to their several niches in nature be explained save by seeing in it the final results of generations of gradually inherited adaptations?

Darwin also believed in the inheritance of acquired characters, although he differed from Lamarck with respect to how such characters are acquired.

Francis Galton in 1875 was one of the first to express skepticism regarding this generally accepted belief but the man who, in a masterly manner, focused the growing doubt, and who did more than any other to inspire thought and investigation upon the subject, was August Weismann, for nearly fifty years professor in the University of Freiburg in Baden. Weismann made the issue so clear that the heritability of acquired characters became the parting of the ways which divided biologists into the two camps of *Neo-Lamarckians* who affirm, and *Neo-Darwinians* who deny, such inheritance. If the question could be decided by a vote or by an expression of opinion, the result would be doubtful, since each column contains the names of men whose scientific accomplishments entitle them to a respectful hearing. Geneticists and embryologists, representing the two lines of study which furnish the most immediate approach to this problem, are well-nigh agreed, however, that acquired characters are not inherited.

But just what are the facts of the case?

5. Confusion in Definitions

The source of much of the lack of agreement in this controversy lies in the definition of what constitutes an "acquired character." One is reminded of the two knights who fought so bitterly over the

color of a shield, one maintaining that it was red,
the other that it was black. So they hacked away
at each other, as all good knights should do in the
defense of the truth, until they both fell down dead
beside the shield which was black on one side and
red on the other.

Of course actual characters are never inherited, but
only the determiners or potentialities which regulate
the way in which the organism reacts to its environment or training with respect to the characters in
question. Reid has pointed out that in one sense
every adult character is "acquired" because it has
no expression at first. For instance, there is no
beard on the face of a male infant, but it will presumably be "acquired" later on in the life-cycle.

It is plain that every new character which represents a forward evolutionary step must have been
"acquired," or added, sometime and somewhere, else
it would not be present, as there is evidence that it is.
Perhaps the question, as Montgomery has suggested,
ought to be changed to read: "*What kinds of acquired
characters are inherited?*" It is obvious that discussion is futile until a common denominator in the
shape of a definition of acquired characters shall be
accepted.

6. WEISMANN'S CONCEPTION OF ACQUIRED CHARACTERS

Weismann defines an acquired character as *any
somatic modification that does not have its origin in the
germplasm.*

Of course those somatic modifications which are phases of the developing individual, such as the acquisition of a deeper voice at puberty or the substitution of the permanent dentition for the milk-teeth, are somatic variations which have their rise and control in the germplasm and consequently cannot properly be included under the head of acquired characters.

Examples of acquired characters in the Weismannian sense are mutilations, the effects of environment, the results of function as in the use or disuse of certain organs, and such diseases as may be due either to invading bacteria or to the neglect or abuse of the bodily mechanism.

7. THE DISTINCTION BETWEEN GERMINAL AND SOMATIC CHARACTERS

Redfield has recently thrown light on the classification of the characters which make up the individual by quoting the familiar lines: —

> "Some are *born* great,
> Some *achieve* greatness,
> Some have greatness *thrust* upon them."

"Born" characters are constitutional, having their origin in the germplasm itself. They are never Weismannian acquired characters and may be illustrated by eye-color, mental disposition, or facial features. Lightning calculators and musical prodigies may have their gifts developed and enlarged, but the fact that their talent is nevertheless an unmistakable gift, and not an acquisition, remains true.

"Achieved" characters are functional and are gained by exercise. Many things are achieved, however, which are not acquired characters, as, for instance, wealth, reputation, or an education. Not any of these are biological characters, and therefore we are not concerned with them in this connection, although in the case of education it should be noticed that the mental exercise necessary to bring about a trained mind, if not the subject matter of the education itself, is distinctly an acquired character of the "achieved" type.

"Thrust" characters are the results of environment. They are acquired without functional activity on the part of the organism and usually in spite of anything the organism can do to prevent. Sometimes these characters are thrust upon the individual before birth, as in the case of blindness caused by parental gonorrhœa or tuberculosis arising from uterine infection, in which case they are termed *congenital* characters.

Congenital or prenatal characters, however, are in no way the same as germinal characters, for they fall just as truly into the category of acquired variations as do those which make their appearance in later life.

8. What Variations reappear?

Returning now to Montgomery's question, — "What kinds of acquired characters are inherited?" — it

Only character acquired from the germplasm can be inherited.

is apparent that only the "born" ones can be, which have their roots in the germplasm whence the new individual arises, and that "achievements" and "thrusts," in order to reappear in the succeeding generation, can do so *only by first becoming incorporated in the germplasm.*

Any character that is not acquired must have been present in the germplasm from which the organism arose, as there is no transfer of characters between organisms except through the germ-cells. Thus it is evident that the only inherited acquisitions are those which, either primarily or secondarily, bring about variation in the germplasm. Such temporary acquisitions as a coat of tan or a display of freckles do not impress the germplasm, for when the cause that incites their appearance is removed, they soon vanish.

9. WHAT MAY CAUSE GERMPLASM TO VARY OR TO ACQUIRE NEW CHARACTERS?

The causes which bring about changes in the germplasm may be either internal or external.

Of possible internal causes may be mentioned first the "amphimixis" of Weismann, that is, the mixture of two nearly related strains of germplasm in sexual reproduction within a species, or secondly, the mixture of two more remotely related strains resulting in hybridization. In either case the strain of germplasm undergoes a shake-up that may result at least in new combinations of characters, if not in the production of entirely new characters. This recombination of

G

characters in amphimixis and hybridization will receive further attention in a later chapter.

The fact that successive parthenogenetic generations, in which amphimixis does not of course occur, may show a larger degree of variability than sexually produced generations, indicates that amphimixis in itself is by no means sufficient to account for all kinds of variations.

The abrupt way, for instance, in which mutations appear in apparent independence of external influences suggests that there may be some internal factor, as yet unknown, acting directly through the germplasm, regardless of external causes.

The assumption of an unknown factor does not necessarily imply a return to "vitalism," which is so elusive of experimental test and hence so unsatisfactory to the scientific mind, nor does it admit, simply because this factor is at present an unknown quantity, that it is consequently doomed to remain so.

It is easily conceivable that the external factors acting upon the germplasm may be grouped into two alternative classes: first, external factors that act upon the somatoplasm and through the agency of the somatoplasm affect the germplasm; and second, those that act directly upon the germplasm without necessarily at the same time influencing the somatoplasm.

The first category, that of somatic modifications which leave their impress upon the germplasm, includes true acquired characters according to our definition, while the second, which includes cases of the direct influence of external stimuli upon the germplasm, regardless of any simultaneous modification of the somatoplasm, must be excluded as irrelevant to a discussion of the heritability of acquired characters in the Weismannian sense, since they are not somatic modifications at all.

Many instances of direct influence of external stimuli upon germplasm are known in biological literature, and these have led to some of the misunderstandings concerning the "interminable question" of the inheritance of acquired characters.

MacDougal, for example, was able by injecting certain salts into the carpels of plants to stimulate the germplasm of the forming seeds so directly that a progeny of modified character was produced which, in succeeding generations, apparently bred true to the newly induced character. This is an instance of what has been termed "parallel induction," where somatoplasm and germplasm are affected together by an external factor, as opposed to "somatic induction" or Weismannian acquired characters, in which the germplasm is secondarily influenced *through*, or *by the agency of*, the somatoplasm. Sitkowski fed the caterpillars of the moth *Tineola biselliella* with an aniline dye (Sudan red III), obtaining therefrom, instead of the normal whitish ones, moths that laid colored eggs, and these in turn hatched into caterpillars still tinged with the color of the red dye. Riddle, with guinea-pigs, and Gage, with poultry, obtained quite similar results. This apparent case of parallel induction, however, is not a matter of inheritance at all, but of animals who got their red color while they were eggs within the mother's body.

10. Weismann's Reasons for doubting the Inheritance of Acquired Characters

Weismann's reasons for questioning the popularly accepted view that acquired characters are inherited may be briefly stated as follows: —

First, there is no known mechanism whereby somatic characters may be transferred to the germ-cells.

Second, the evidence that such a transfer actually does occur is inconclusive and unsatisfactory.

Third, the theory of the continuity of the germ-plasm is sufficient to account for the facts of heredity without assuming the inheritance of acquired somatic characters.

Let us examine these three statements a little more closely.

11. No Known Mechanism for impressing the Germplasm with Somatic Characters

The somatoplasm is something that has traveled out from the original fundamental germplasm along the paths of differentiation and elaboration. The more complex the body cells become, that is, the more successive modifications they undergo, the more difficult it is for these somatic cells to return to their original primitive estate.

In many lower forms of life where cell elaboration is not so great, a part lost by amputation is often regenerated, but this process is not possible in higher forms where the parts represent cell complexes too hopelessly differentiated to begin anew the unfolding sequences of their elaboration. This difficulty was a very real one in the mind of that famous nocturnal inquirer Nicodemus when he asked: "How can a man be born when he is old? Can he enter a second time into his mother's womb and be born?"

Not only the development of the race which we call evolution, but also the determination of the individual in heredity, is a *chain of onward-moving sequences* like the succession of events in history. It is hard to see how recent events can influence preceding events. It is hard to see how the water that has gone over the dam can return and affect the flow of the river upstream in any direct way. It is likewise hard to see how differentiated somatoplasm, which represents the end stage of a successive series of modifications, can make any definite impress upon the original germplasmal sources from which it arose.

Darwin felt this difficulty and presented with apologies his provisional hypothesis of *pangenesis* in which he assumed that every bodily part sends contributions to the germ-cells in the form of "gemmules." These gemmules, or hypothetical somatic delegates, then reconstruct in the germ-cells the characters of the entire body, including acquired modifications as well as all others, and thus there is no reason why acquired characters cannot readily be transmitted. Unfortunately there is no tangible basis in fact for this delightfully simple explanation to rest upon. It is a theory assuming that *all parental somatic cells* take part in the formation of the new individual, hence it was called "pangenesis," or *origin from all.*

Nothing we have subsequently learned of minute cell structure favors this hypothesis, while many facts go quite against it. Moreover, it is directly opposed to the theory of the continuity of germplasm so convincingly set forth later on by Weismann. Darwin indeed advanced it only in the most tentative way, being entirely ready to see it abandoned at any time for something better. It at least performed one valuable service to science, namely, that of demonstrating how far investigators were from an adequate conception of any means by which somatic modifications might become incorporated in the germ-cells.

We must acknowledge, however, with Lloyd Morgan that the fact that a mechanism for the transfer of somatic characters to the germ-cells has not been discovered, is not proof that such a mechanism does not exist. It may simply be beyond our present powers of penetration.

12. Evidence for Transmission of Acquired Characters Inconclusive

The evidence for the inheritance of acquired characters was, for a long time, taken for granted. This theory was the most obvious explanation of many facts and so was accepted without question. An obvious interpretation, however, is not always the correct one. The sun appears to go around the earth, but astronomers assure us that it does not.

When Weismann began to sift the evidence for the inheritance of acquired characters, he found that it was largely based upon opinion rather than fact, much like the popular belief with regard to prenatal influences and birthmarks, or the causation of warts by handling toads.

The supposed evidence for the inheritance of acquired characters falls chiefly into four categories : —

a. Mutilations;
b. Environmental effects;
c. The effects of use or disuse;
d. The transmission of disease.

a. Mutilations

It is fortunate that the sons of warriors do not inherit their fathers' honorable scars of battle, else we would now be a race of cripples.

The feet of Chinese women of certain classes have for centuries been mutilated into deformity by bandaging, without the mutilation in any way becoming an inherited character. The same result is also true of circumcision, a mutilation practised from ancient times by the Jews and certain other Eastern peoples. The progressive degeneration or crippling of the little toe in man has been explained as the inheritance of the cramping effect of shoes upon generations of shoe wearers, but, as Wiedersheim has pointed out, the fact that Egyptian mummies show the same crippling of the little toe is unfavorable to this hypothesis, for no ancient Egyptian could

ever be accused of wearing shoes or of having had
shoe-wearing ancestors.

Sheep and horses with docked tails as well as dogs
with trimmed ears never produce young having the
parental deformity. Weismann's classic experiment
with mice, an experiment subsequently confirmed
by others, is additional negative evidence upon this
same point.

What Weismann did was to breed mice whose
tails had been cut off short at birth. He continued
this decaudalization through twenty-two generations
with absolutely no effect upon the tail-length of the
new-born mice. One may see in the catacombs of
the Zoologisches Institut at Freiburg, filed carefully
away on shelves, as a "document," long rows of labeled
bottles containing the fifteen hundred and ninety-two
martyrs to science which made up the twenty-two
generations of mice in this famous experiment.

Conklin has hit the nail upon the head with respect
to mutilations by saying: "Wooden legs are not inherited, but wooden heads are."

b. Environmental Effects

Trees deformed by prevailing winds, like the
willows that line the canals in Belgium and Holland,
or storm-crippled trees along the exposed seacoast
are not known to produce a modified progeny
when their adverse environmental conditions are
removed. Similarly, the persistent sunburn of Eng-
lishmen long resident in India does not reappear in
their children born in England.

Sumner kept mice in a constant but abnormally high temperature of 26° C. with the result that the ears, tail, and feet grew noticeably larger than in control animals kept in ordinary lower temperatures, while at the same time the general hairiness of the body decreased. It should be remembered, however, that mice are mammals which pass through an extended uterine existence, so that it is easy to see how the offspring in this case were subjected to the same excessive temperature as the parents for a period sufficient to amply account for their subsequent variation when removed to a normal environment.

Zederbaur finds that the wayside weed *Capsella*, which in the course of many years has gradually crept along the roadside up into an Alpine habitat and there "acquired" Alpine characters, upon being transplanted to the lowlands retains its Alpine modifications. Although this case has been cited as an authentic instance of the inheritance of acquired characters, is it not possible that the conquest of the Alps by *Capsella* has been due, in the course of time, not to the inheritance of acquired characters at all, but to a gradual natural selection of just those germinal variations which best fitted it to cope with Alpine conditions until, finally, a strain of germplasm producing somatoplasm suitable to Alpine conditions has been isolated in the form of an elementary species derived from the original type? If this is what has happened, of course such germplasm would give rise to Alpine plants whether individual plants grew to maturity near the snow-line or in the warm valleys at a lower altitude.

Marie von Chauvin was able, by decreasing the amount of water in an aquarium, to transform the gill-breathing salamander *Axolotl* into the land form, *Amblystoma*, which in its adult form has no gills, but breathes by means of lungs. Both of these forms are sexually mature, reproducing their like, and had long been recognized by systematists as distinct species.

More recently Kammerer, by similarly reducing the water supply, succeeded in transforming *Salamandra maculosa*, a salamander that normally produces about seventy eggs which, when hatched in water, become gill-breathing tadpoles, into a salamander producing only two to seven young which are born alive without gills and are able to live out of water entirely, in damp situations. These land-adapted offspring, moreover, when supplied with abundant water, produce in turn tadpoles which spend days only, instead of months, in the water undergoing their metamorphosis, thus showing an apparent inheritance of an acquired character.

It should be pointed out, however, that in these cases the gill-breathing forms in each instance represent a case of arrested development. *Axolotl* is simply a larval form of *Amblystoma* that, under normal conditions of an abundant water environment and high temperature, gets no further in its metamorphosis than the tadpole stage, when it produces eggs and sperms and finishes its life story. A change in environment simply permits the life-cycle to go on further. Changing from gill-breathing to lung-breathing is not, therefore, an acquired character, but a purely germinal character that may be either blocked or released by changing conditions in the environment.

c. The Effects of Use or Disuse

The callosities on the end of a violinist's left-hand fingers are acquired by use, but they are not inherited. There are callosities on the knees of the wart-hog, *Phacechœrus*, which are also apparently the result of use, for these animals kneel as they root for a living in the African forests, and have done so for untold generations. It has been noticed that young wart-hogs as soon as they are born possess the callosities, so that this instance looks like one of inheritance of a character acquired through use or exercise.

The skin on the soles of human feet is thicker than the skin elsewhere, and by use it becomes still thicker. This is apparently another instance of the same sort. The writer has observed, however, that a cross section through the foot of a "mud puppy," *Necturus maculatus*, shows a much thickened sole. *Necturus*, it should be noted, is a very primitive salamander living always under water and never using the soles of its feet in any way to bear its weight, nor is it reasonable to suppose that it ever had any ancestors who did so, for the hands and feet of the Amphibia are the most primitive and ancient hands and feet to be found in the animal kingdom without any known ancestral types. The thickening of the skin on the sole of the mud puppy's feet must be due, therefore, to germinal determiners and is in no way an acquisition through use. The same may also be true of the wart-hog's knees and of human soles.

The strong arm, the skilled hand, and the trained ear are not inherited. They have always to be reacquired in each succeeding generation just as surely as the ability to walk, or to read and write.

Herbert Spencer has defined instinct as "inherited habit." But surely those instincts which determine a single isolated action during the lifetime of the individual, such as the spinning of a peculiar cocoon, cannot be the result of habit, since habits are formed only through repeated action. If, then, *some* instincts require a different explanation from that of "inherited habit," may it not be likely that all instincts do? Dr. Hodge, who succeeded in hatching tame quail chicks out of "wild" eggs, asks the pertinent question: "How can a *fear* hatch out of an egg?" The habit of wildness, particularly with precocial chicks like quails, may, under an inciting environment, be very soon established, but it is difficult to see how caution, gained by the experience of the parents, can find its way into the fertilized egg.

d. Disease Transmission

Many diseases, like tuberculosis, have their immediate cause in invading pathogenic bacteria. Bacteria themselves cannot be inherited for the reason that it is not possible for them to become an integral part of the fertilized egg and thus cross the "hereditary bridge" which joins two generations. A general predisposition to bacterial disease, that is, a lack of resistance to bacterial invasion due to defectiveness in physical or physiological equipment, may be present as a combination of characters in the germplasm, or an individual, as the result of disease, may "acquire" a generally weakened germplasm and so produce a progeny exhibiting general liability to disease; but it is doubtful if such a condition can properly be termed the inheritance of an acquired character, since the particular definite disease in question is not demonstrably heritable.

When alcoholism "runs in a family," its reappearance in the son is probably due to the fact that he is derived from the same weak strain of germplasm as his father. The fact that the father succumbed to the alcohol habit is not the determining cause of drunkenness in the son. The same thing that caused the father to become an alcoholic, namely, weak germplasm, and not the resulting drunkenness in the parent, is the causal factor for alcoholism in the son.

At the same time it is entirely probable that hereditary alcoholism may in some cases arise through "parallel induction," that is to say, acquired alcoholism may end in the simultaneous poisoning and consequent modification of both the somatoplasm and germplasm of the parent, with the result that the germplasm has less resistance to alcoholism in a succeeding generation. The offspring are consequently more likely to succumb to the disease. This, how-

ever, is not the inheritance of an acquired character
or of a definite somatic modification.

When a man of the present generation has rheu-
matic gout, it is a severe stretch both of patriotism
and of the powers of heredity to trace the origin of
the affliction back to a revolutionary ancestor who
acquired sciatic rheumatism by sleeping on the
ground at Valley Forge, yet this is quite as direct as
many alleged instances of the inheritance of disease.

In the majority of instances, apparent cases of
disease inheritance are merely instances of *reinfection*. This reinfection of the offspring may occur
very early in embryonic life, or it may happen after
birth, provided the offspring are exposed to the same
environment as that in which the parent acquired the
disease, but in any case *reinfection is not heredity*.

13. THE GERMPLASM THEORY SUFFICIENT TO AC-
COUNT FOR THE FACTS OF HEREDITY

Weismann holds that the theory of the continuity
of the germplasm, already considered in a previous
chapter, is sufficient in itself to account for the facts of
heredity. Hence it is quite unnecessary to fall back
upon the inheritance of acquired characters as an explanation, since this theory is at least difficult, if not
impossible, of satisfactory proof.

To prove the inheritance of acquired characters,
according to Weismann three things are necessary:
first, a particular somatic character must be called
forth by a known external cause; *second*, it must be
something new or different from what was already

exhibited before, and not be simply the reawakening of a latent germinal character; and *third*, the same particular character must reappear in succeeding generations in the absence of the original external cause which brought the character in question forth. As yet these conditions have not been convincingly met in the evidence which has been brought forward in support of the inheritance of acquired characters.

14. The Opposition to Weismann

The opponents of Weismann point out, as a weak place in his argument, the assumption that the germ-plasm is so insulated from the somatoplasm as not to be influenced by it. Weismann assumes, of course, that the germplasm is isolated from the somatoplasm very early in the development of the fertilized egg into an individual, and that when once isolated it thereafter takes no active part in, nor is in any way affected by, the vicissitudes through which the somatoplasm, or the body itself, passes. The somatoplasm is thus merely a carrier of the germplasm and unable to affect the character of it any more than a rubber hot-water bag, although capable of assuming a variety of shapes, can affect the character of the water within it.

In opposition to this view it is urged that every organism is a physiological as well as a morphological unity, and that cells entirely insulated within such a unity would be a physiological miracle.

There is abundant evidence that germ-cells, or rather the sexual organs producing the germ-cells, do affect the somatoplasm under particular conditions, as, for example, in cases of castration when those somatic features called "secondary sexual characters" undergo profound modification.

If the germplasm thus exercises a constant influence on the somatoplasm, why, it seems legitimate to ask, may not the reverse be true and acquired somatic characters leave their impress upon the germ-cells?

15. CONCLUSION

But even granting the reverse to be true, that is, that the somatoplasm affects the germ-cells, the inheritance of acquired characters is by no means thereby established.

In order to do this, the precise acquired character in question, which indirectly exercised its influence upon the germ, must be redeveloped, and, although the germplasm might conceivably receive an influence from the somatoplasm and be affected by it in a general way, it is a different matter entirely to develop anew the verisimilitude of the character itself which is supposed to have been acquired.

It will be seen in subsequent pages, under the discussion of data furnished by experimental breeding, that the weight of probability is decidedly against the time-honored belief in the inheritance of acquired characters.

CHAPTER VI

THE PURE LINE

1. THE UNIT CHARACTER METHOD OF ATTACK

In reducing any body of facts to a science, it is first necessary to determine the underlying units out of which the facts are made up.

Chemistry was alchemy until the chemical elements were identified and isolated. Histology was *terra obscura* until the cell theory brought forward "cells" as the units of tissues. In the same way there could be no science of genetics until the conception was developed that the individual is a bundle of unit characters rather than a unit in itself. So it has come about that we now speak of inheritance as applied to unit characters rather than to individuals as a whole.

Incidentally the fact that an organism is a combination of many units makes it easy to account for the wide diversity of forms found in nature, since the addition of a single unit greatly increases the possible combinations in successive generations.

Thus if three unit characters, *A*, *B*, and *C*, are present in each parent, for example, there would be six possible double combinations in the offspring, namely, *AA*, *AB*, *AC*, *BB*, *BC*, and *CC*. If now a fourth unit D is added by one parent, there would be not only the original six double combinations, but in addition to these, AD, BD, and CD, that is, as many more as there are unit characters with which the new one may combine.

Obviously, when individuals are made up of very many unit characters, as, for instance, a thousand, the addition of one new unit character will increase the possible double combinations a thousand fold.

2. Galton's Law of Regression

Galton was one of the first [1] to attempt to express mathematically the relationship between parents and offspring by means of treating statistically a single unit character. According to Galton, a mathematical expression of the relationship between two generations should serve as a corner-stone of heredity.

What Galton did was to take human stature as a unit character in comparing 204 English parents and their 928 adult offspring, because human stature is not complicated by environmental influences and is, consequently, a purely hereditary matter.

Since female height is normally less than male height, the two were reduced, for purposes of comparison, to a common male denominator by multiplying each female height by 1.08, which is the average amount that the male exceeds the female in height. There are always two parents concerned in the sexual production of every offspring,

---

[1] "Hereditary Genius," 1869.

therefore Galton reckoned a "midparent" in each case, according to the formula

$$\frac{1.00 \; \delta + 1.08 \; \female}{2}$$

in order to represent the double parental generation by a single number for the purpose of easy comparison with the filial generation.

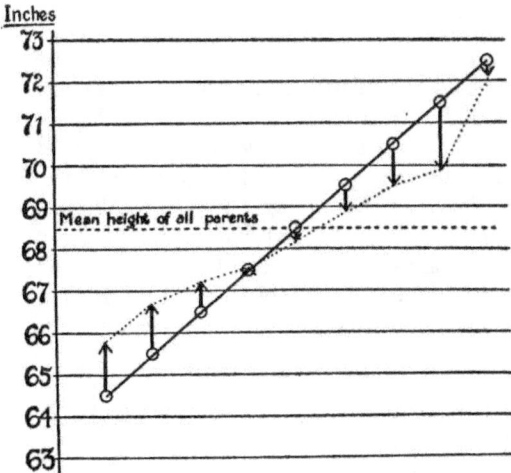

FIG. 33. — Scheme to illustrate *Galton's law of regression*. The circles represent graded groups of parental height while the arrowpoints indicate the average heights attained by the respective offspring. The offspring of undersized parents are taller, and of oversized parents are shorter than their respective parents. Based on data from Galton.

The results of his measurements expressed in inches are shown in the following table, in which the

offspring are, in each instance, arranged under their
respective midparents.

Midparental height . .	64.5	65.5	66.5	67.5	68.5	69.5	70.5	71.5	72.5
Average height of offspring .	65.8	66.7	67.2	67.6	68.3	68.9	69.5	69.9	72.2

The mean group for all the midparents, it will be
seen, is 68.5, and the offspring of this group average
68.3.   The table is expressed graphically in Figure
33 in which the circles connected by the diagonal
line represent the graded parental heights, while the
arrowpoints indicate the average heights of the offspring in each group.   In order to compare these
two series of numbers more readily, they may be
reduced to a common basis in which the mean class
in each instance is made equal to 100, as follows: —

Midparental height . .	94	95.5	97	98.5	100	101.5	103	104.5	106
Average height of offspring .	96	97.5	98.5	99	100	101	101.5	102	105.5

The same series may be expressed in terms of
amount of deviation from the mean or middle classes,
as shown below.   The deviation of each group in the
series is marked by the signs + or − according as
the heights given are greater or less than 100.

Midparental height . .	− 6	−4.5	−3	−1.5	0	+ 1.5	+ 3	+ 4.5	+ 6
Average height of offspring .	− 4	−2.5	−1.5	−1	0	+ 1	+ 1.5	+ 2	+ 5.5

Finally, the relation between the midparent and the average offspring may be expressed in fractional form by taking the average height of the offspring for the numerator and the height of the midparent for the denominator in each instance. The minus deviations are thus seen to be

$$\frac{4}{6} + \frac{2.5}{4.5} + \frac{1.5}{3} + \frac{1}{1.5}$$

which, added together, divided by four and reduced to a decimal, equal .60.

Similarly, the plus deviations are

$$\frac{1}{1.5} + \frac{1.5}{3} + \frac{2}{4.5} + \frac{5.5}{6}$$

which reduce to .63. The average of the minus deviations (.60) and the plus deviations (.63) is nearly .62, or about two thirds.

That is to say, the fraction $\frac{2}{3}$ represents the amount of resemblance or "inheritance" between two generations, as determined by the foregoing series, while the remaining $\frac{1}{3}$ is the measure of "regression" from the general type.

This illustrates Galton's *Law of Regression* or the tendency in successive generations toward mediocrity. The law may be stated as follows: —

Average parents tend to produce average children; minus parents tend to produce minus children; plus parents tend to produce plus children; *but the progeny of extreme parents, whether plus or minus, inherit the parental peculiarities in a less marked degree than the latter were manifested in the parents themselves.*

3. THE IDEA OF THE PURE LINE

It was Galton's law of regression that suggested to the Danish botanist Johannsen a possible means of controlling heredity. In his mind arose the question whether it would not be possible by continually breeding from plus parents, granting that plus parents produce plus offspring and making allowance for some regression to type, to shove over the offspring more and more into the plus territory and so to establish a plus race.

To test this hypothesis, Johannsen selected beans, *Phaseolus*, with which to experiment, since this group of plants is self-fertilizing, prolific, and easily measurable. Somewhat to his surprise, his beans refused to shove over as much as expected. That is, big beans did not yield principally big offspring, nor little beans little offspring, according to the expectation, although they each produced offspring that varied in the manner of fluctuating variability around an average unlike the parental type. This gave Johannsen the idea that he was using mixed material, so he next isolated the progeny of single beans, which, being self-fertilized, each constituted unmistakably a single hereditary line. In this way nineteen beans, now famous, became the known ancestors of Johannsen's original nineteen "pure lines," a further study of which has led the way to some of the most brilliant biological discoveries of recent years.

A pure line has been defined by Johannsen as "the descendants from a single homozygous organism exclusively propagating by self-fertilization," and more briefly by Jennings as "all the progeny of a single self-fertilized individual."

4. JOHANNSEN'S NINETEEN BEANS

It was found by Johannsen that the progeny of each of these pure lines of beans varied around its own mean, which was different in each of the nineteen instances. When, however, extremes from any pure line series were selected and bred from, the progeny, instead of showing two thirds inheritance and one third regression with respect to the extremeness of a particular character, as Galton found was true in the case of human stature, showed *no inheritance and complete regression* away from the extreme condition of the parent bean back to the type for the entire pure line in question. That is, *selection within a pure line is absolutely without effect* in modifying a particular character in the offspring of the line in question.

This is illustrated in Figure 34 in which the results of selecting for size in the year 1902 is shown for four pure lines only. The average for each pure line is given at the top of its column. When, for example, beans weighing 60 cg. were selected from pure lines II, VII, and XV, the average weights of their progeny were 56.5, 48.2, and 45.0 cg. respectively, which in each instance is nearer to the average for the pure line than to the weight of the parental seed.

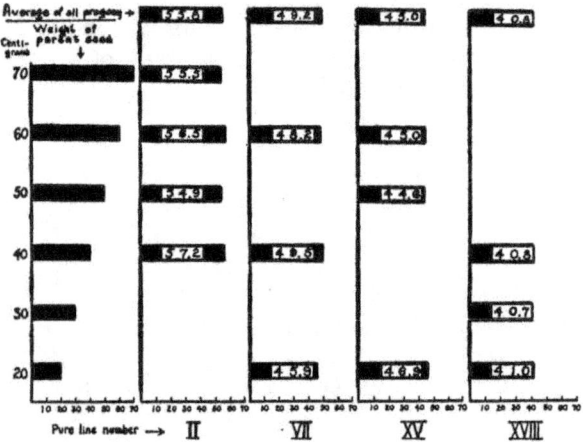

FIG. 34. — The result of selection in four pure lines of beans. The vertical columns, representing the average progeny from different sized parents all derived from the same pure line, contain groups nearer alike than the horizontal columns, representing progeny from the same sized parents, but different pure lines. All the numbers indicate centigrams. Data from Johannsen.

It will be seen at once that the averages in the vertical columns are nearer alike than the averages in the horizontal columns. In other words, the beans bred true to their pure line rather than to their fluctuating parent.

As a further example of this law, take the result

of selection for six years in pure line I as shown in the accompanying table and in Figure 35.

HARVEST YEAR	MEAN WEIGHT OF SELECTED PARENT SEED		MEAN WEIGHT OF OFFSPRING	
	Minus	Plus	From Minus Parent	From Plus Parent
1902	60	70	63.15	64.85
1903	55	80	75.19	70.88
1904	50	87	54.59	56.68
1905	43	73	63.55	63.64
1906	46	84	74.38	73.00
1907	56	81	69.07	67.66

It is evident, for instance, that in 1907 the smallest beans, weighing an average of 56 cg., gave an average progeny weighing 69.07 cg, while the largest ones for the same year, weighing an average of 81 cg., produced nearly the same average in their progeny as did the smallest beans, that is, 67.66 cg.

Incidentally all the progeny from both large and small parents averaged notably less in 1904 than all the progeny from large and small parents in 1906, a result due to a "poor year" when certain factors of environment were unfavorable. Such unfavorable conditions, however, are known to influence in no way the hereditary qualities of the beans. Thus it appears that, although the progeny of a pure line present plenty of variations of the fluctuating type, due probably to environmental differences in nutrition, moisture, etc., such variations are quite inef-

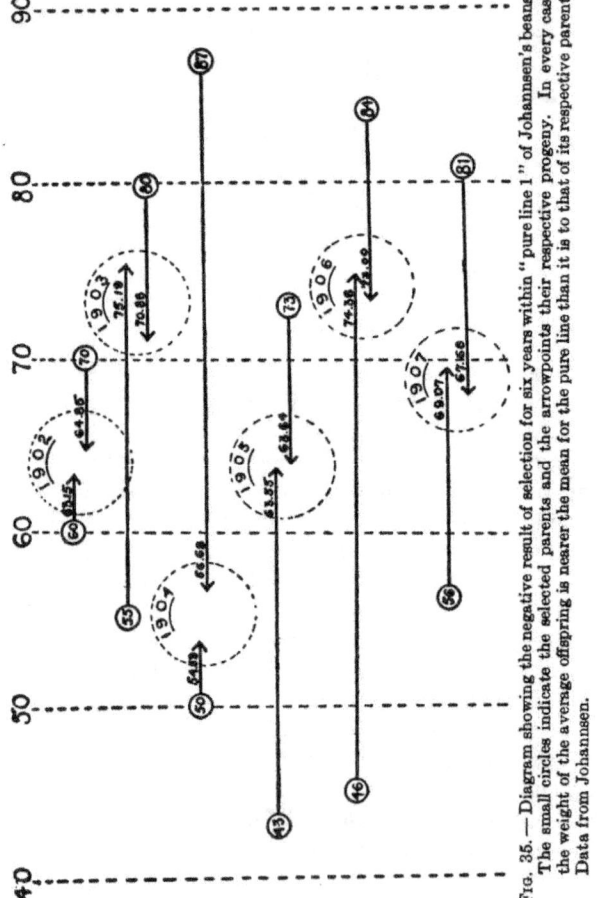

Fig. 35. — Diagram showing the negative result of selection for six years within "pure line 1" of Johannsen's beans. The small circles indicate the selected parents and the arrowpoints their respective progeny. In every case the weight of the average offspring is nearer the mean for the pure line than it is to that of its respective parent. Data from Johannsen.

fectual so far as inheritance is concerned, and it makes no difference whether the largest or the smallest beans within a pure line are selected from which to breed, the result will be the same, in that there is a complete return to mediocrity or type with no "inheritance" of the parental modification. As a matter of fact in 1903, 1906 and 1907 the lighter parents gave a heavier progeny than the heavier parents.

It will be seen at once that here is a discovery of far-reaching importance which may require us to reconstruct certain cherished ideas about the part played in the evolution of species, as well as in heredity, by natural selection.

5. CASES SIMILAR TO JOHANNSEN'S PURE LINES

Although according to Johannsen pure lines are "the progeny of a single self-fertilized individual," it is plain that in at least three other possible cases something quite similar to "pure lines" may be obtained.

First, when two similar organisms identical in their germinal determiners with regard to a particular character inbreed, their progeny will form a pure line *so far as this particular character is concerned* just as truly as two parents that are united in a single individual produce a pure line progeny as the result of self-fertilization.

Second, in cases of parthenogenesis, the progeny arising from a single female individual without the customary qualitative reduction of chromosomes that accompanies sexual reproduction, constitute a pure line or an unmixed strain.

Third, in cases of asexual reproduction where the progeny are simply the result of continued fission of the original individual, a pure line may be said to continue from generation to generation.

In the second and third categories it should be pointed out that the "pure line" is assured only so long as asexual reproduction continues. It is quite possible for an organism, heterozygotic in composition, to continue to breed true or to produce an apparently pure line so long as asexual methods are employed. As soon as such an organism, however, changes to the sexual method of reproduction, segregation of characters may occur and different combinations result.

6. Tower's Potato-beetles

As an illustration of the effect of selection within pure lines of the first category may be mentioned a case given by Tower in his exhaustive experiments on the Colorado potato-beetle *Leptinotarsa decemlineata*. Among the numerous cultures of this beetle which were under control, a considerable variation in color made its appearance. For convenience in classification these variations were graded into arbitrary classes or *graduated variants* (see p. 52) ranging from dark to light.

When a male and a female from the extreme class at the dark end of the series were allowed to breed together, their progeny were not dark, but fluctuated

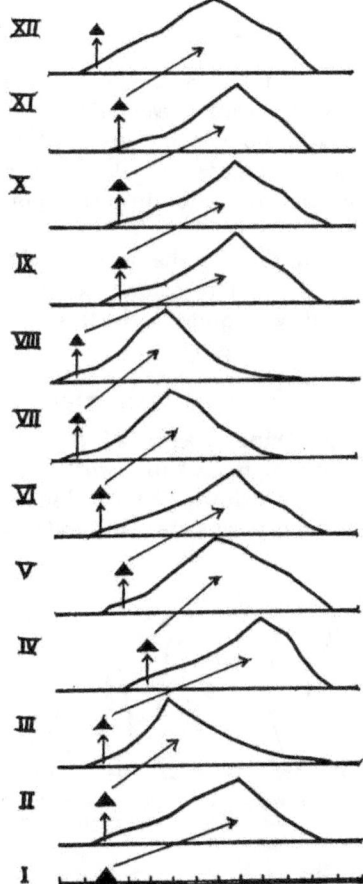

FIG. 36. — Diagram showing the ineffectiveness of selection through twelve generations within a homozygous strain in the case of the Colorado potato-beetle (*Leptinotarsa*). In each generation extreme dark specimens were selected as the parents of the succeeding generation but the progeny always swung back to the type. After Tower.

in color around the original average of the entire series. This process of selecting each time an extreme pair of dark parents was continued for twelve generations, as shown in Figure 36, without in any way increasing the percentage of brunette potato beetles in the progeny.

Thus in a pure line formed by the breeding of two individuals alike with respect to color, the selection of an extreme variant was quite without effect in modifying the color of the progeny.

7. Jennings' Work on Paramecium

An instance of the third category of pure lines is furnished by Jennings' remarkable work on the protozoan *Paramecium*, which was published in 1909. Jennings carried on his experiments quite independ-

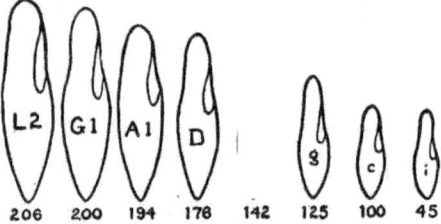

| L2 | G1 | A1 | D | | g | c | i |
| 206 | 200 | 194 | 178 | 142 | 125 | 100 | 45 |

Fig. 37.—Eight pure races of *Paramecium*. The actual mean length of each race is given in micra below the corresponding outline. Magnified about 230 diameters. After Jennings.

ently of Johannsen, but he nevertheless arrived at the same general conclusion, namely, that selection within a pure line is without effect.

Jennings found that Paramecia differ from each

other in size, structure, physical character, and rate of multiplication as well as in the environmental conditions required for their existence and, furthermore, that these differences, in an hereditary sense, are "as rigid as iron."

With respect to the character of mean length he was able to isolate eight races, or pure lines, whose average size, drawn to scale, is shown in Figure 37.

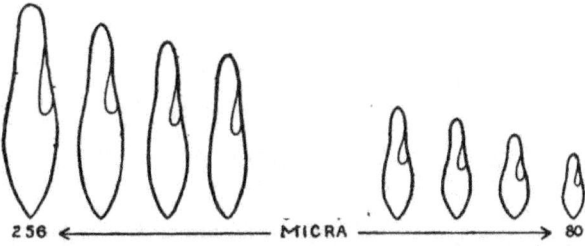

256 ← —————————— MICRA —————————— → 80

Fig. 38. — Diagram of a single race (*D*) showing the variation in the size of the individuals. Magnified about 230 diameters. After Jennings.

Each of these pure lines produced a progeny which exhibited a considerable range of fluctuating variation. The offspring of *pure line D*, for example, varied from 256 to 80 micra [1] in length with an average of 176 micra, as shown in Figure 38, where samples of the different classes of variants in *pure line D* are arranged in a series.

A single representative of each of the different classes of variants out of all of the eight pure lines bred by Jennings is shown in Figure 39.

Each horizontal row represents a single race or

———

[1] A micron is $\frac{1}{1000}$th of a millimeter.

pure line, the average size of which is indicated by the
sign +. The mean length of the entire lot, as shown

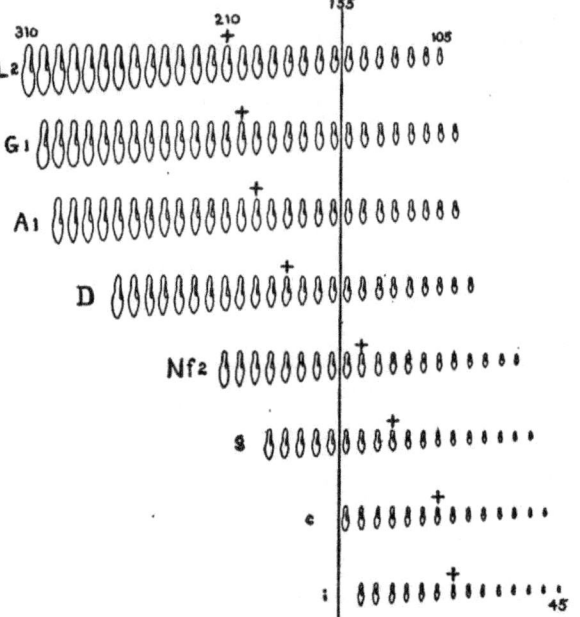

FIG. 39.—Diagram of the species *Paramecium* as made up of the eight
different races shown in Figure 37. Each horizontal row represents a
single race. The individual showing the mean size in each race is indicated by a cross placed above it. The mean for the entire lot is at
the horizontal line. The magnification is about 24 diameters. After
Jennings.

by the vertical line, is 155 micra. The total number
of individuals belonging to each size is not indicated,
but in every horizontal line their number is more

numerous near the average for that line and less numerous at the extremes, thus forming the typical normal frequency polygons of fluctuating variability.

The significant fact about these series is this, that extreme individuals selected from any pure line do not reproduce extreme sizes like themselves, but instead, a progeny varying according to the laws of chance around the average standard of the particular line from which it came.

8. PHENOTYPICAL AND GENOTYPICAL DISTINCTIONS

From the foregoing it will be seen that the behavior of an organism in heredity cannot always be determined by an inspection of its somatic characters alone.

For example, six Paramecia, each 155 micra in length and apparently identical, could be selected from the six upper pure lines in Jennings' table given in Figure 39 which would produce six progenies *definitely unlike*, whereas in the case of *pure line D*, twenty-four Paramecia, all measurably different from each other in size, would be found to produce twenty-four progenies *practically identical*.

Organisms that appear to be alike, regardless of their germinal constitution, are said by Johanssen to be identical *phenotypically*, or to belong to the same *phenotype*.

On the other hand, organisms having identical germinal determiners such as those of the varying members of *pure line D*, are said to be *genotypically* alike or to belong to the same *genotype*.

I

Organisms belong to the same phenotype with respect to any character when their somatoplasms are alike. They belong to the same genotype when their germplasms are alike.

The word "genotype" was suggested by Johannsen in honor of Darwin and his theory of pan*gene*sis, although there are certain objections to its use in this connection for the reason that systematists have already appropriated it in a different sense.

Natural history and common usage deal principally with phenotypes, that is, with organisms as they appear. The older theories of heredity were likewise concerned with phenotypes, but we are now coming to see more clearly than before that heredity must always be a case of similarity in origin, that is, in germinal composition, and that similarity in appearance by no means always indicates similarity in origin or true relationship.

The assumption that similarity in appearance does indicate relationship has been made the foundation of many conclusions in comparative anatomy and phylogeny, but to the modern student of genetics who places his faith in *things as they are,* rather than *in things as they seem to be,* conclusions based upon phenotypical distinctions alone have in them a large source of error which must be taken into account.

In a museum of heredity, should such a collection ever be assembled, the specimens would not be arranged phenotypically as they are in an ordinary museum where things that look alike are placed together as if in bonds of relationship, but they would be arranged *historically* from a genetic point of view to show their true origin one from another.

9. The Distinction between a Population and a Pure Line

A mixture of pure lines has been called a *population*.

It is not possible to distinguish a pure line from a population by inspection, since both may be phenotypically alike. Fluctuations about the average occur in both cases with no appreciable difference in character, although such fluctuations, when they occur within a pure line, are simply somatic differences caused in general probably by modifications in nutrition or some other external factor of environment, while fluctuations in a population include not only modifications of this transient nature, but also permanent hereditary differences due to germinal differences in the various pure lines of which the population is composed.

Johannsen has made the distinction between pure lines and populations clear by the following figure (Fig. 40), in which five pure lines of beans are combined artificially to form a population.

The beans which make up the pure lines noted in this figure are represented inclosed within inverted test tubes. The beans in any single tube are all of one size. Tubes vertically superimposed upon each other also contain only beans of one size.

Thus it is seen that what may be a rare size of bean in one line, for instance that in the left-hand tube of *pure line 3*, may be identical with the commonest size in another line, as *pure line 2*. The five pure lines represented in Figure 40 are combined in a *population* at the bottom of the figure, making a phenotype that marks the five phenotypes above, which are also five genotypes. In the population, however, the five genotypes are hidden within one phenotype.

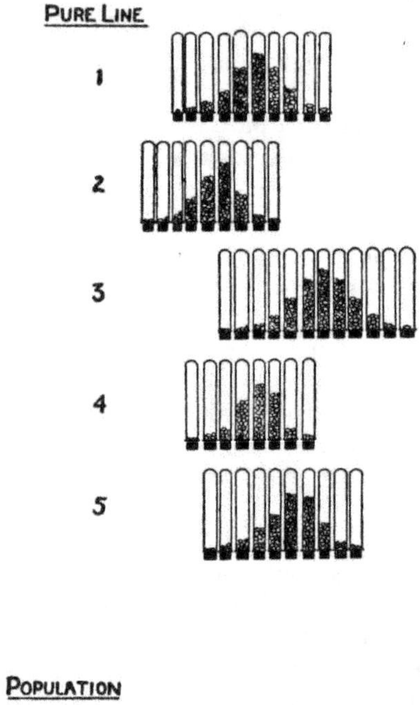

PURE LINE

1

2

3

4

5

POPULATION

Fig. 40. — Diagrams showing five *pure lines* and a *population* formed by their union. The beans of each pure line are represented as assorted into inverted test tubes making a curve of fluctuating variability. Test tubes containing beans of the same weight are placed in the same vertical row. After Johannsen.

Hence, while selection within a pure line has no hereditary influence, it is evident that selection within a population may shift or move over the type of the progeny obtained, in the direction of the selection simply by isolating out a pure line of one type. Thus beans chosen from the extreme left-hand test tube in the population cited would belong only to *pure line 2*, while those taken from the extreme right-hand test tube could belong only to *pure line 3*.

Galton's "law of regression," namely, that minus parents give minus offspring and plus parents plus offspring, with a tendency to reversion from generation to generation, depends simply upon a partial but not complete isolation of pure lines out of a population.

From this distinction between pure lines and populations it is clear why breeders in selecting for a particular character out of their stock need to keep on selecting continually in order to maintain a certain standard. As soon as they cease this vigilance, there is a "reversion to type" or, as they say, "the strain runs out," which means that the pure lines become lost in the mixed population which inevitably results as soon as selective isolation of the pure line ceases.

Such reversion must always be the case in dealing with a population made up of a mixture of pure lines, for only by the isolation of pure lines can the constancy of a character be maintained. When, however, a pure line is once isolated, then all the members of it, large as well as small, are equally efficient in maintaining the pure line in question, regardless of their phenotypical constitutions.

10. Pure Lines and Natural Selection

From the foregoing statements it appears that by means of selection within a population, such as occurs normally in nature, it is not possible to get anything out that was not already there to begin with. If this is so, the origin of species cannot have come about, as Darwin thought, through natural selection by a gradual accumulation of slight favorable variations. The best that selection can do is to isolate pure lines. Within pure lines it is quite powerless to change the genotypical characters. In other words, natural selection can only maintain and strengthen the frontier posts that are already established. It cannot break into the wilderness and create new centers.

Since the extreme members of a pure line, having the same genotypical constitution, always tend to backslide to mediocrity within the limits of the line in question, the crucial question is: How can the critical step from one genotype to another, a step indispensable in the evolutionary derivation of species, ever occur? That it has repeatedly occurred in the course of time is amply proven by the fact that somehow or other we have gone from Ameba to man.

At present the only loophole of escape seems to lie either in the unlikely inheritance of acquired characters, or in mutations which make the leap from one character to another, and so eventually from one type to another, without the aid of selection.

It is interesting to note that Johannsen himself, who has been so prominently concerned in erecting this barrier in the way of the evolutionary derivation of species by natural selection, has recently reported mutations arising within his pure lines of beans. It must be admitted that to the skeptical there is a vicious circle here, for when a variation fails to reappear in a subsequent generation, it may be explained as the failure of natural selection to act within a pure line, but when a variation *does* reappear it is hailed as a mutation!

In any event the way of experiment lies open, and the evidence of investigators in this critical field will be awaited with keen interest.

CHAPTER VII

SEGREGATION AND DOMINANCE

1. METHODS OF STUDYING HEREDITY

MODERN studies in heredity have been pursued principally in three directions: first, by microscopical examination of the germ-cells; second, by statistical consideration of data bearing upon heredity; and third, by experimental breeding of animals and plants.

The first two of these methods of approach have already been touched upon as well as experimental breeding with reference to "pure lines." In the present chapter attention will be directed to a consideration of experimental breeding with reference to hybridization, that is, breeding from unlike parents, a process which Jennings characterizes by the expressive phrase, "the melting-pot of cross-breeding."

2. THE MELTING-POT OF CROSS-BREEDING

Hybridization, or cross-breeding, as formulated by Galton (1888), results in one of three kinds of inheritance, namely, blending, alternative, or particulate.

Of these, *blending inheritance* may be called the typical "melting-pot" in which contributions from the two parents fuse into something intermediate and different from that which was present in either parent. Galton illustrated this process by the inheritance of human stature in which a tall and a short parent produce offspring intermediate in height. A more thorough consideration of this type of inheritance will be presented in Chapter IX.

By the method of *alternative inheritance* the parental contributions do not melt upon union, but retain their individuality, reappearing intact in the offspring. In inheritance of human eye-color, for example, the offspring usually have eyes colored like those of one of the parents when the parental eye-color is unlike in the two cases, rather than eyes intermediate in color between those of both parents.

According to Galton *particulate inheritance* results when the offspring present a mosaic of the parental characters, that is, when parts of both the maternal and paternal characters reappear in the offspring without losing their identities by blending or without excluding one another. Piebald races of mice arising from parents with solid but different colors have been cited as illustrations of this sort of inheritance, although it will be seen later in connection with the "factor hypothesis" that another interpretation of this phenomenon is not only possible but probable.

The distinctions between these three categories of inheritance are diagrammatically represented in Figure 41.

In blending inheritance the offspring are seen to be unlike either parent, because the parental determiners fuse into a new thing. In alternative inheritance, on the contrary, the offspring may be like either parent, since the characters in question do not lose their individuality upon union, as shown in the diagram. Only one or the other of the two

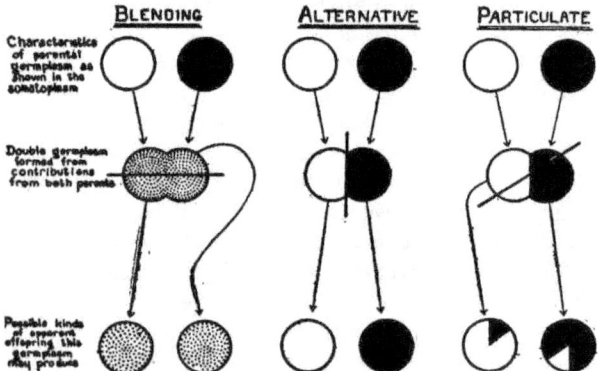

Fig. 41. — Three kinds of inheritance described by Galton.

mutually exclusive characters thus becomes effective in determining the nature of each offspring.

Finally, in particulate inheritance the double germplasm which determines a new individual may be imagined to undergo a diagonal rather than a vertical cleavage upon maturation, thereby causing unblended fragments of both parental characters to become effective at once, in this manner producing a mosaic offspring.

3. JOHANN GREGOR MENDEL

Our understanding of the working of inheritance in hybridization we owe largely to the unpretentious studies of an Austrian monk, Johann Gregor Mendel, who, although a contemporary of Darwin, was probably unknown to him. For eight years Mendel carried on original experiments by breeding peas in the privacy of his cloister garden at Brünn and then sent the results of his work to a former teacher, the celebrated Karl Nägeli, of the University of Vienna. At the time Nägeli's head was full of other matters, so that he failed to see the significance of his old pupil's efforts. However, in 1866 Mendel's results appeared in the Transactions of the Natural History Society of Brünn,[1] an obscure publication that reached hardly more than a local public. Here Mendel's investigations were buried, so to speak, because the time was not ripe for a general appreciation or evaluation of his work.

At that time neither the chromosome theory nor the germplasm theory had been formulated. Moreover, much of our present knowledge of cell structure and behavior was not even in existence. Weismann had not yet led out the biological children of Israel through the wilderness upon that notable pilgrimage of fruitful controversy which occupied the last two decades of the nineteenth century, and the attention of the entire thinking world was being monopolized

---

[1] Verhandlungen naturf. Verein in Brünn. Abhandl. IV, 1865 (which appeared in 1866).

by the newly published epoch-making work of Charles Darwin.

Mendel died in 1884, and his work slumbered on until it was independently discovered almost simultaneously by three botanists whose researches had been leading up to conclusions very much like his own. These three men were de Vries of Holland, von Tschermak of Austria, and Correns of Germany. Their papers were published only a few months apart in 1900 and were closely followed by important papers from Bateson in England and Davenport and Castle in America, with a rapidly increasing number from other biologists the world over. To-day the literature upon this subject has grown to be very large, and the end is by no means yet in sight.

Concerning Mendel, Castle has well said: "Mendel had an analytical mind of the first order which enabled him to plan and carry through successfully the most original and instructive series of studies in heredity ever executed."

4. Mendel's Experiments on Garden Peas

What Mendel did was to hybridize certain varieties of garden peas and keep an exact record of all the progeny, in itself a simple process but one that had never been faithfully carried out by any one.

Before examining Mendel's results it may be well to state the difference between normal and artificial self-fertilization. Self-fertilization occurs when from the pollen and ovule of the same flower are derived the two gametes which uniting produce a zygote that develops into the seed and subsequently into the adult plant of the next generation.   In artificially crossing normally self-fertilized flowers it is necessary to carefully remove the stamens from one flower while its pollen is still immature, and later, at the proper time, to transfer to it ripe pollen from another flower.

Mendel's cross-breeding experiments on peas showed certain numerical relations which gave rise to what has come to be rather indefinitely known as "Mendel's law."   This law may be temporarily formulated as follows : —

When parents that are unlike with respect to any character are crossed, the progeny of the first generation will apparently be like one of the parents with respect to the character in question.   The parent which impresses its character upon the offspring in this manner is called the *dominant*.   When, however, the hybrid offspring of this first generation are in turn crossed with each other, they will produce a mixed progeny, 25 per cent of which will be like the dominant grandparent, 25 per cent like the other grandparent, and 50 per cent like the parents resembling the dominant grandparent.

An illustration will serve to make plain the manner in which this law works out.

Mendel found that when peas of a tall variety were artificially crossed with those of a dwarf variety, all the resulting offspring were tall like the first parent.   It made no difference which parent was selected as the tall one. The result was the same in either case, showing that the character of tallness is independent of the character for sex.

When these tall cross-bred offspring were subsequently crossed with each other, or allowed to produce offspring by self-fertilization which amounts to the same thing, 787 plants of the tall variety and 277 of the dwarf kind were obtained, making approximately the proportion of 3 to 1.

On further breeding the dwarf peas thus derived proved to be pure, producing only dwarf peas, while the tall ones were of two kinds, one third of them "pure," breeding true like their tall grandparent, and two thirds of them "hybrid," giving in turn the proportion of three tall to one dwarf like their parents.

These crosses may be expressed as follows: —

$$\text{Tall, } T, \times \text{ dwarf, } t, = \text{tall, } T(t).$$

That is, tallness crossed with dwarfness equals tallness with the dwarf character present but latent.

Mendel termed the character, which became apparent in such a hybrid, in this case tallness, the *dominant*, and the latent character which receded from view, in this instance dwarfness, the *recessive*.

When now the hybrids, $T(t)$, were crossed together, the result algebraically expressed was as follows: —

$$
\begin{array}{l}
T + t \ \text{(all possible egg characters)} \\
\underline{T + t} \ \text{(all possible sperm characters)} \\
TT + \quad Tt \\
\quad\quad \underline{Tt \quad + tt} \\
TT + 2\,T(t) + tt
\end{array}
$$

That is, one out of four possible cases was dwarf, *t*, in character and the other three were apparently tall, although only one out of the three was pure tall, *T*, while the remaining two were tall with the dwarf character latent, *T* (*t*).

The same thing may be expressed more graphically by the checkerboard plan, which Punnett suggested (Fig. 42). Each square of the checkerboard represents a zygote which, having received a gamete from each of the two parents, may develop into a possible offspring. The character of the gametes of the parents is shown outside of these squares, while the arrows represent the parental source

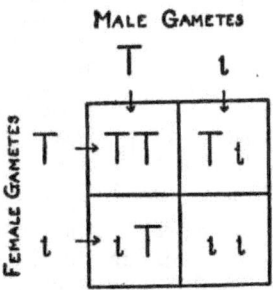

Fig. 42. — Diagram to illustrate theoretically the formation of the four possible zygotes in the second filial generation of a monohybrid.

from which the offspring have received their hereditary composition.

The essential feature of Mendel's law is briefly this: *hereditary characters are usually independent units which segregate out upon crossing, regardless of temporary dominance.*

Mendel carried on further experiments with garden peas, using other characters. He obtained practically the same result as in the instance already given, for the actual progeny in the second generation of the cross-bred offspring figured up, as seen in the table below, very nearly to the expected theoretical ratio of 3 to 1.

Character	Number of Dominants	Number of Recessives	Ratio
Form of seed	5474 smooth	1850 wrinkled	2.96 to 1
Color of seed coat . .	6022 yellow	2001 green	3.01 to 1
Color of flowers . . .	705 colored	224 white	3.15 to 1
Form of pods	882 inflated	299 constricted	2.95 to 1
Color of unripe pods . .	428 green	152 yellow	2.82 to 1
Position of flowers . .	651 axial	207 terminal	3.14 to 1
Length of vine . . .	787 tall	277 dwarf	2.84 to 1
Total	. . . . . .	. . . . . .	2.98 to 1

Darbishire repeated the yellow-green cross with garden peas, obtaining in the second generation the large total of 139,837 individuals of which 105,045 were yellow and 34,792 green, which is very close to 3 to 1.

5. Some Further Instances of "Mendel's Law"

Since the rediscovery of Mendel's law the ratio of 3 to 1 in the second generation has been found by a number of different investigators to be constant in a large array of characters observed both in animals and plants of diverse kinds when these are cross-bred with reference to the characters in question.

Botanists have the advantage perhaps in this matter, as they deal with forms which usually produce a large number of offspring from a single cross, — a very desirable thing in estimating ratios. On the other hand, they are handicapped by being unable usually to obtain more than one generation in a year, while zoologists may secure from many animals like rabbits and mice several generations in a year, although ordinarily the number of progeny is much

ORGANISM	AUTHOR	DOMINANT	RECESSIVE	DATE
Nettles	Correns	Serrated leaves	Smooth-margined leaves	'03
Sunflower	Shull	Branched habit	Unbranched habit	'08
Cotton	Balls	Colored lint	White lint	'07
Snapdragon	Baur	Red flowers	Non-red flowers	'10
Wheat	Biffen	Susceptibility to rust	Immunity to rust	'05
Tomato	Price and Drinkard	Two-celled fruit	Many-celled fruit	'08
Maize	de Vries	Round, starchy kernel	Wrinkled, sugary kernel	'00
Silkworm	Toyama	Yellow cocoon	White cocoon	'06
Cattle	Spillman	Hornlessness	Horns	'06
Pomace fly	Morgan	Red eyes	White eyes	'10
Horses	Bateson	Trotting habit	Pacing habit	'07
Land snail	Lang	Unbanded shell	Banded shell	'09
Mice	Darbishire	Normal habit	Waltzing habit	'02
Guinea-pig	Castle	Short hair	Angora hair	'03
Canaries	Bateson and Saunders	Crest	Plain head	'02
Poultry	Davenport	Rumplessness	Long tail	'06
Man	Farrabee	Brachydactyly	Normal joints	'05
Barley	von Tschermak	Beardlessness	Beardedness	'01
Salamander (Amblystoma)	Haecker	Dark color	Light color	'08

smaller and the ratios obtained have a larger chance of error than is the case with the more numerous plant offspring.

Semi-microscopic animals, as, for example, the pomace fly, *Drosophila*, which produces a large progeny every two weeks or so, may combine the general advantages mentioned for the two groups of organisms

K

indicated above, but they have the disadvantage of being so small that the detection of their distinctive phenotypic characters is attended with considerable technical difficulty.

What the modern experimenter in genetics desires is an organism, first, that possesses conspicuous distinctive somatic characters, and, second, which will come to sexual maturity early and breed either in captivity or under cultivation both numerously and frequently.

The preceding table, compiled chiefly from Bateson [1] and Baur,[2] might easily be much extended. It shows from what diverse sources confirmatory evidence of the truth of Mendel's law has been derived within the past few years.

6. THE PRINCIPLE OF SEGREGATION

The essential thing which Mendel demonstrated was the fact that, in certain cases at least, the determiners for heredity derived from diverse parental sources may unite in a common stream of germplasm from which, in subsequent generations, they may segregate out apparently unmodified by having been intimately associated with each other. This "law of segregation" depends upon the conception that the individual is made up of a bundle of unit characters. It may be illustrated by the separate flowers picked from a garden which, after being made into a nosegay, may be taken apart and rearranged without in

[1] "Mendel's Principles of Heredity," 1909.
[2] "Einführung in die experimentelle Vererbungslehre," 1911.

any way disturbing the identity of the separate blossoms.

The general formula of segregation that covers all cases of organisms cross-bred with respect to a single character, that is, *monohybrids*, is given in Figure 43.

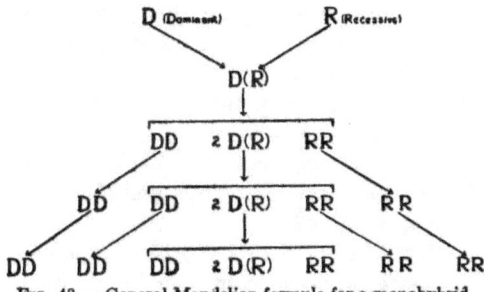

Fig. 43. — General Mendelian formula for a monohybrid.

7. Homozygotes and Heterozygotes

A character which is present in the offspring in double quantity because it was present in both parents is said by Bateson to be *homozygous*, while an organism which is homozygous with respect to any character is called a *homozygote* so far as that particular character is concerned.

In contrast to the homozygous condition, an organism is said to be *heterozygous* when it derives the determiner of a character from one parent only. Such an organism is described as a *heterozygote* with respect to the character in question. A homozygous and a heterozygous dominant may appear alike, although not necessarily so, that is, they may have the same phenotypical constitution, but their genotypical composition is always different.

8. The Identification of a Heterozygote

"Homozygote" and "heterozygote" are terms then descriptive solely of the genotypical constitution of organisms, and, as has been said, it is not always possible to distinguish one from the other by inspection, although it may frequently be done, as will be pointed out later. *The only sure way to identify a heterozygote is by breeding to a recessive and observing the kind of offspring produced.*

Peas of the formulæ TT and $T(t)$, for example, both look alike, since a single determiner for the tall character, T, is sufficient to produce complete tallness. When, however, these two kinds of tall peas are each bred to a recessive dwarf pea, of the formula tt, the progeny will differ distinctly in the two cases as follows : —

Case I. $T + T \times t + t = 100$ per cent $T(t)$.

Case II. $T + t \times t + t = 50$ per cent $T(t) + 50$ per cent tt.

That is, if the dominant to be tested is homozygous (Case I), the entire progeny will exhibit the dominant character, but if the dominant to be tested is heterozygous (Case II), then only one half of the progeny will show the character in question.

9. The Presence and Absence Hypothesis

Mendel's conception that every dominant character is paired with a recessive alternative is now being largely replaced by *the presence and absence hypothesis* which was first suggested by Correns but later logically worked out by others, particularly by Hurst, Bateson, and Shull. According to this latter interpretation, a determiner for any character either is, or is not, present. When it is present in two parents, then the offspring receive a double, or *duplex*, "dose," to use Bateson's word, of the determiner. When it is present in one parent only, then the offspring have a single, or *simplex*, dose of the character. When it is present in neither parent, it follows that it will not appear in the offspring. In this case the offspring are said to be *nulliplex* with respect to the character in question. Take the case of tall and dwarf peas, the determiner for tallness when present produces tall peas, even if it comes from one parent only, but if this determiner for tallness is absent from both parents, the offspring are nulliplex, that is, the absence of tallness results and only dwarf peas are produced.

The difference between the presence and absence theory and the dominant and recessive theory is that in the former case the "recessive" character has no existence at all, while in the latter instance it is present, but in a latent condition.

10. DIHYBRIDS

So far reference has been made exclusively to *monohybrids*, any two of which are supposed to be similar except with respect to a single unit character. Monohybrids are comparatively simple, but when two organisms are crossed which differ from each other with respect to *two* different unit characters, the situation becomes more complicated.

Mendel solved the problem of dihybrids by crossing wrinkled-green peas with smooth-yellow peas. He found that *smoothness* S is dominant over *wrinkledness* W and that *yellow color* Y is dominant over *green* G, or, as it would be stated according to the presence and absence theory, smoothness is a positive character which fills out the seed-coat to plumpness while its absence leaves a wrinkled coat, and yellowness is a positive character due to a fading of the green which causes the yellow to be apparent. In the absence of this green fading factor or determiner the green, of course, appears.

If smooth-yellow SY and wrinkled-green WG are crossed, all the offspring are smooth-yellow, but they carry concealed the recessive determiners for wrinkledness and greenness according to the formula $S(W)Y(G)$. When the determiners of these cross-breds segregate out during the maturation of the germ-cells, they may recombine so as to form four possible double gametes, namely, smooth-yellow SY and wrinkled-green WG, which are exactly like the grandparental determiners from which they arose, and in addition, two entirely new combinations, smooth-green SG and wrinkled-yellow WY.

Since the male and the female cross-breds are each furnished with these four possible gametic combinations, the possible number of zygotes formed by their union will be sixteen $(4 \times 4 = 16)$. That is, the monohybrid proportion of 3 to 1 in dihybrid combinations is squared, $(3+1)^2 = 16$.

It of course does not follow that the offspring in dihybrid crosses will always be sixteen in number, or that they will always conform strictly to the theoretical expectation of $(3+1)^2$. The offspring obtained undoubtedly obey the laws of chance, but the greater the number of offspring, the nearer they come to falling into the expected grouping.

The sixteen possible zygotes resulting from a dihybrid cross will give rise to sixteen possible kinds of individuals which in turn, as will be demonstrated directly, present four kinds of phenotypic and nine kinds of genotypic constitutions.

A dihybrid mating, using the same symbols employed in the case just described, would be expressed algebraically as follows : —

$$
\begin{array}{llll}
SG+ & WY+ & SY+ & WG = \text{all the possible egg gametes} \\
SG+ & WY+ & SY+ & WG = \text{all the possible sperm gametes} \\
\hline
SGSG+ & SGWY+ & SGSY+ & SGWG \\
& SGWY & & +WYWY+ \ WYSY+ \ WYWG \\
& & SGSY & + \ WYSY \qquad +SYSY+ \ SYWG \\
& & SGWG & \qquad + \ WYWG \qquad + \ SYWG+WGWG \\
\hline
\end{array}
$$
$$SGSG+2\,SGWY+2\,SGSY+2\,SGWG+WYWY+2\,WYSY+2\,WYWG+SYSY+2\,SYWG+WGWG$$

The second and the ninth items in this result are alike; by combining them the revised result reads : —

$$SGSG+4\,SGWY+2\,SGSY+2\,SGWG+WYWY+2\,WYSY+2\,WYWG+SYSY+WGWG$$

There are then these nine different combinations of germinal characters or nine different genotypes in any dihybrid cross. By placing the recessive characters in parentheses, whenever the corresponding dominant is present to indicate that the dominant causes the former to recede from view, these nine genotypes may be combined into four phenotypes as follows : —

Phenotypes . .	9 SY	3 SG	3 WY	1 WG
Genotypes . .	4 $S(G)(W)Y$ 2 $S(G)SY$ 2 $SY(W)Y$ $SYSY$	$SGSG$ 2 $SG(W)G$	$WYWY$ 2 $WYW(G)$	$WGWG$

From this analysis it may be said that the Mendelian ratio for a typical dihybrid is phenotypically 9 : 3 : 3 : 1, while that for a monohybrid, as we have already seen,

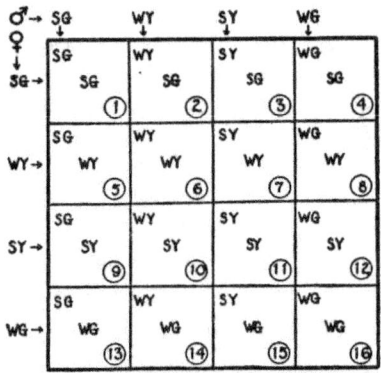

Fig. 44. — Diagram to illustrate the possible combinations arising in the second filial generation (F_2) following a cross between yellow-smooth YS and green-wrinkled GW peas.

is 3 : 1. This expected ratio corresponds essentially with the actual results Mendel obtained in crossing smooth-yellow and wrinkled-green peas.

Figure 44 presents a graphic representation of the different combinations resulting from a dihybrid cross following the checkerboard plan used in Figure 42 to illustrate monohybrids.

The nine genotypes and four phenotypes which result from a dihybrid cross are shown in the following squares.

Number in Each Class	GENOTYPE	Number of Squares in Fig. 44	PHENOTYPE	Number in Each Class
1	$SYSY$	11		
2	$(W)YSY$	7·10	SY	9
2	$S(G)SY$	3·9		
4	$S(G)(W)Y$	2·5·12·15		
1	$SGSG$	1	SG	3
2	$SG(W)G$	13·4		
1	$WYWY$	6	WY	3
2	$WYW(G)$	8·14		
1	$WGWG$	16	WG	1
16				16

Another illustration of dihybridism is shown in Figures 45 and 46 which is based upon data furnished by the Davenports.[1] In the matings given here, dark or pigmented hair, represented by the solid black circles, is dominant over light-colored, that is, unpigmented or slightly pigmented hair, symbolized by the open circles, while curly hair is dominant

[1] "Heredity of Eye-color in Man," *Science*, N. S. 26, p. 589, 1907; "Heredity of Hair Form in Man," *Amer. Nat.* 42, p. 341, 1908. Davenport, C. B. and G. C.

over straight, represented by crooked and straight lines respectively in the diagram. In other words, the presence of pigment is dominant over the absence of pigment, while the factor that causes curliness is dominant over the absence of this factor, with respect to human hair.

HOMOZYG.

XS

FIG. 45. — The heredity of human hair according to data by C. B. and G. C. Davenport. The arcs represent the somatoplasms of four individuals. Within the arcs are the gametes formed by these individuals. The dominant character is placed on the outside of the arc where it will be visible.

When a homozygous individual with dark curly hair crosses with a homozygous individual with light straight hair, all the offspring have dark curly hair.

The dark curly-haired individuals of this second generation, however, are heterozygous with respect to each of these two hair characters. When any two individuals having this particular genotypic composition mate, therefore, they may produce any one of four possible phenotypes — dark curly, dark straight, light curly or light straight haired individuals. These four phenotypes in turn will present nine different genotypic combinations out of sixteen possible cases, as shown in Figure 46.

Figure 45 furthermore serves to make clear, *first*, the distinction between somatoplasm and germplasm; *second*, the maturation of germ-cells; *third*, the

Number in each class	Genotype	Phenotype	Number in each
4			
2		Dark curly	9
2			
1			
1		Dark straight	3
2			
1		Light curly	3
2			
1		Light straight	1
16			16

Fig. 46. — Diagrams showing the possible genotypic and phenotypic combinations resulting when two heterozygous individuals, with dark curly hair, mate. Symbols are the same as in Figure 45.

segregation of gametes; and *fourth*, the formation of zygotes in sexual reproduction.

The cells of the somatoplasm are represented as

making up the arcs within which are inclosed the germ-cells after their reduction through maturation, which results in giving to each germ-cell half the number of determiners that are present in the somatic cells.

It will be remembered that when two gametes, or mature germ-cells, unite, they form a zygote having the proper number of determiners normal to the species in question instead of double that number. Symbols for dominant characters in the diagram are placed on the *outside* of the somatic arcs, because these are the characters that are visible or phenotypic, while the non-apparent recessives are placed on the inside out of sight.

11. The Case of the Trihybrid

Mendel went even further and computed the possibilities which would result when two parents were crossed differing from each other with respect to three unit characters. He found that the results actually obtained by breeding closely approximated the theoretical expectation.

This expectation in the case of a trihybrid cross is that the cross-breds resulting will all exhibit the three dominant characters, while their genotypic constitution will include six factors, namely, these three dominant characters plus their corresponding recessives or "absences."

Cross-breds of the first generation will, therefore, have eight possible kinds of triple gametes and when interbred may form a possible range of sixty-four

(8×8) different zygotes, which corresponds to a monohybrid raised to the third power $(3 + 1)^3$. These sixty-four zygotes group together in eight

	♂→	RSP	R·P	RSp	R·p	rSP	r·P	rSp	r·p
♀↓		↓	↓	↓	↓	↓	↓	↓	↓
RSP→		RSP RSP	R·P RSP	RSp RSP	R·p RSP	rSP RSP	r·P RSP	rSp RSP	r·p RSP
R·P→		RSP R·P	R·P R·P	RSp R·P	R·p R·P	rSP R·P	r·P R·P	rSp R·P	r·p R·P
RSp→		RSP RSp	R·P RSp	RSp RSp	R·p RSp	rSP RSp	r·P RSp	rSp RSp	r·p RSp
R·p→		RSP R·p	R·P R·p	RSp R·p	R·p R·p	rSP R·p	r·P R·p	rSp R·p	r·p R·p
rSP→		RSP rSP	R·P rSP	RSp rSP	R·p rSP	rSP rSP	r·P rSP	rSp rSP	r·p rSP
r·P→		RSP r·P	R·P r·P	RSp r·P	R·p r·P	rSP r·P	r·P r·P	rSp r·P	r·p r·P
rSp→		RSP rSp	R·P rSp	RSp rSp	R·p rSp	rSP rSp	r·P rSp	rSp rSp	r·p rSp
r·p→		RSP r·p	R·P r·p	RSp r·p	R·p r·p	rSP r·p	r·P r·p	rSp r·p	r·p r·p

Fig. 47. — Diagram showing the possible combinations in a guinea-pig trihybrid of the F_2 generation. R, rosetted coat ; r, non-rosetted coat (absence of R) ; S, short hair ; s, angora hair (absence of S) ; P, pigmented ; p, albino (absence of pigment). The eight possible triple gametes of each parent are placed in the upper and left hand margins. Each of the sixty-four squares represents a possible zygote or fertilised egg, having received a triple gamete from each parent.

different phenotypes and twenty-seven different genotypes.

The trihybrid cross with its resulting combinations is well illustrated by Castle's work on guinea-pigs which confirms the Mendelian hypothesis on an

Number in each class	Genotype	Phenotype	Number in each class
1	SS PP RR		
2	SS Pp RR		
2	Ss PP RR		
4	Ss Pp RR	SPR	27
2	SS PP Rr	Short, pigmented, rosetted	
4	SS Pp Rr		
4	Ss PP Rr		
8	Ss Pp Rr		
1	SS pp RR		
2	Ss pp RR	SpR	9
2	SS pp Rr	Short, albino, rosetted	
4	Ss pp Rr		
1	ss PP RR		
2	ss Pp RR	sPR	9
2	ss PP Rr	Angora, pigmented, rosetted	
4	ss Pp Rr		
1	SS PP rr		
2	SS Pp rr	SPr	9
2	Ss PP rr	Short, pigmented, non-rosetted	
4	Ss Pp rr		
1	ss pp RR	spR	3
2	ss pp Rr	Angora, albino, rosetted	
1	SS pp rr	Spr	3
2	Ss pp rr	Short, albino, non-rosetted	
1	ss PP rr	sPr	3
2	ss Pp rr	Angora, pigmented, non-rosetted	
1	ss pp rr	spr	1
		Angora, albino, non-rosetted	
64			64

extensive scale.   In Figure 47 dominant characters are represented by capital letters, while recessives or absences are indicated by corresponding small letters.

When a smooth, or non-rosetted (r), short-haired (S), pigmented (P) guinea-pig is crossed with a rosetted (R), long-haired (s), albino (p) guinea-pig, all the offspring appear to be of one phenotypic constitution, namely, rosetted, short-haired, and pigmented (RSP).   Their genotypic constitution is represented by the formula $RrSsPp$.   These six factors may form eight possible triple gametes, as follows: $RSP, RsP, RSp, Rsp, rSP, rsP, rsp$.   When two germ-cells each made up of these eight triple gametes unite in sexual reproduction, they will give rise to sixty-four (8×8) possible zygotes as displayed in Figure 47.

An analysis of Figure 47 shows among the offspring eight different phenotypes in the ratio of $27 : 9 : 9 : 9 : 3 : 3 : 3 : 1$ and 27 different genotypes in the proportions indicated on the opposite page.   The *order* of the three pairs of symbols is changed from that in Figure 47 to emphasize the fact that with independent unit characters the order is immaterial.

12. Conclusion

Although the ratios for more than a trihybrid were computed by Mendel, the experimental test has never been carried out, since it involves such large and complicated proportions.

In the case of four differing unit characters in the parental generation, the offspring of the quadruple hybrids derived from such an ancestry would include 256 or $(3 + 1)^4$ possibilities instead of 64 or $(3 + 1)^3$, as in the case of trihybrids. When ten differing characters are combined in the parental generation, there would result over a million possible kinds of offspring among the hybrids of the second generation, $(3 + 1)^{10} = 1,048,576$.

From the foregoing it is apparent that in practical breeding the only hope lies in dealing with not more than one or two characters at a time. Since unit characters usually behave independently of each other, one may breed for a single character until it is segregated out in a homozygous, that is pure, condition, and then in the same way obtain a second character, a third, and so on.

Thus in a few generations of properly directed crosses there can be obtained combinations of characters united in one strain that formerly were never obtained at all or were only hit upon by the merest chance at long intervals. Herein lies the scientific control of heredity which the trinity of Mendelian principles: namely, independent unit characters, segregation, and dominance, has placed in human hands.

13. SUMMARY

Three principles are concerned in Mendel's law: independent unit characters, dominance, and segregation.

a. Independent Unit Characters. An organism, although acting together as a physiological and morphological whole, may be regarded from the point of view of heredity as consisting of a large number of independent heritable unit characters.

b. Dominance. In the germplasm there are certain determiners of unit characters which dominate others during the development of the somatoplasm. In other words, they determine the apparent character of the organism by causing that character to become visible.

The alternative *recessive* characters, although they may be present in the germplasm, are unable to become manifest in the somatoplasm so long as the dominant characters are present. When, however, a dominant character is absent, its recessive alternative becomes manifest.

c. Segregation. Unit characters, although they may be intimately associated together in the individual, during the complicated process of maturation that always precedes the formation of a new individual, separate or segregate out as if independent of each other and thus are enabled to unite into new combinations.

CHAPTER VIII

REVERSION TO OLD TYPES AND THE MAKING OF NEW ONES

1. The Distinction between Reversion and Atavism

THERE are two ways in which types of animals or plants that are different from the present ones may be conceived to arise, namely, by the reappearance of old types and by the formation of new ones. In the reappearance of old types a distinction may be drawn between reversion and what has been termed atavism.

Atavism, or "grandparentism," may be defined as skipping a generation with the result that a particular character in the offspring is unlike the corresponding character in either parent, but instead, resembles the character in one of the grandparents.

In *reversion*, on the contrary, a character reappears which has not been manifest perhaps for many generations, although it was actually present in some remote ancestor. J. Arthur Thomson's definition of reversion is : "All cases where through inheritance there reappears in an individual some character which was not expressed in his immediate lineage, but which had occurred in a remoter, but not hypothetical, ancestor."

This distinction between atavism and reversion becomes clearer by illustration.

If heterozygous brown-eyed individuals mate, there is one possibility in four that their offspring

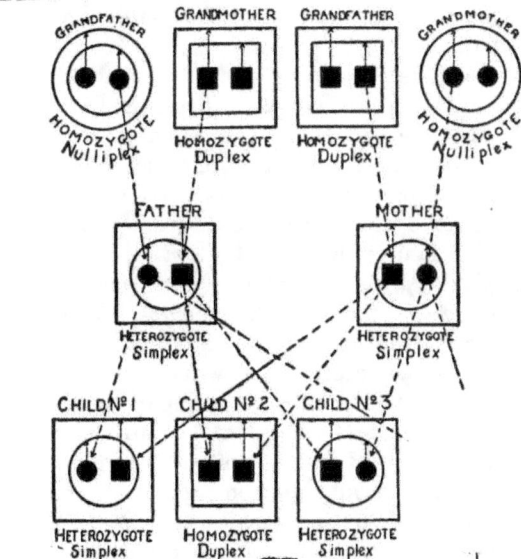

FIG. 48.—Three generations of a Mendelian monohybrid. The outlines represent the somatoplasms with the phenotypic character on the outside. The black symbols inclosed within the somatoplasm stand for the germplasm in the form of gametes. The short dotted arrows indicate the relation between germplasm and somatoplasm. The long dotted arrows indicate possible recombinations of germplasms.

will have blue eyes unlike their own, but like the two blue-eyed grandparents. Such a blue-eyed child would be an instance of atavism. The explanation of this apparently inconsistent hereditary behavior is perfectly simple in the light of the Mendelian ratios, as shown diagrammatically in Figure 48, in which the circles represent the blue-eyed and the squares the brown-eyed character.

This figure also illustrates what typically occurs in the formation of Mendelian monohybrids of the first and second filial generations. The squares are symbols for the *dominant* characters, while the circles are symbols for the *recessive* characters. When the two are superimposed, the circle recedes from view, The large outside figures indicate the *somatoplasm*, therefore the *phenotype*. The small inclosed figures indicate the *germplasm*, therefore the *genotype*. The short dotted arrows indicate what it is that determines the somatoplasm in each case, while the long dotted arrows show what possible recombinations of germplasms can be made. Child No. 4 is an "extracted recessive" derived from dominant parents, but with one recessive grandparent on each side. It is a case of "atavism," or taking after the grandparent. Notice that atavism can occur only by alternative inheritance.

To quote Davenport: "In the majority of cases atavism is a simple reappearance in one fourth of the offspring of the absence of a character due to the simplex nature of the character in both parents."

An illustration of *reversion* would be the reappearance of the ancestral jungle-fowl pattern in domestic poultry or of the slaty blue color of the ancestral rock-pigeon among buff and white domestic pigeons, for the ancestral character or characters in this type of hereditary behavior, as said before, reappear only after a lapse of many generations.

2. FALSE REVERSION

"Around the term ' reversion,' " Bateson observes, "a singular set of false ideas have gathered themselves." In proof of this statement there may be cited at least five categories of apparent reversion which properly ought not to be classed as true reversion.

a. Arrested Development

Feeble-mindedness is not reversion to ancestral forms of less intelligence, but an instance of *arrested development* when, for some reason, the individual fails to accomplish his normal cycle of development.

Likewise harelip in man is not a case of reversion to rabbit-like ancestors in which harelip is the normal condition, but it is ordinarily due to an arrest or failure of certain embryonic steps that are essential to the development of the usual form of human lip.

b. Vestigial Structures

These are the vanishing remains of characters that were formerly of significance. They do not represent something latent that is now *re*appearing, for they have never yet disappeared phylogenetically, and consequently they cannot be regarded as true reversions.

The muscles under the scalp which enable those persons possessing them to wiggle the ears; the palatine ridges in the roof of the mouth of many

babies and some adults which resemble the ridges
in the roof of a cat's mouth; the vermiform appendix, a necessary part of the digestive apparatus of
many animals but fraught so often with evil consequences to man; these and scores of similar characters, which, taken together, make man in the eyes
of the comparative anatomist a veritable old curi-
osity shop of ancestral relics, are the last traces of
characters which formerly had a significance in some
of man's forbears. Having lost their usefulness,
these structures still hang on to the anatomical
household as pensioners. They have not been recalled from the past, but have always been with us,
although of diminishing importance. In no sense,
therefore, can they be called reversions.

c. Acquired Characters resembling Ancestral Ones

Sometimes the drunken descendant of a drunken
great-grandparent has acquired this characteristic
through his own initiative quite aside from any ancestral contribution to his germplasm. This is not
reversion. It is a reacquisition which resembles
the ancestral condition.

Again, tame animals that run wild acquire habits
resembling those of their wild ancestors, but this is
not necessarily reversion. It is the natural response
of feral animals to the conditions of wild life.

d. Convergent Variation

The European hedgehog, *Erinaceus*, an insecti-
vore, the American porcupine, *Erithizon*, a rodent,
and the Australian spiny anteater, *Echidna*, a

monotreme, are all mammals which have developed
in a similar manner the very peculiar device of der-
mal spines.   There is no reason, however, for regarding this character as due to descent from a common
spiny ancestor.   It is not reversion to an ancestral
type, but rather a case of convergent variation.
Similarity does not always indicate genetic continuity.
In the case of birds albinism, melanism and fla-
vism are modifications of ordinary pigmentation which
appear irregularly among many different species as
pathological " sports," but no one of these conditions
can be regarded as reversions to ancestral white,
black, or yellow types.

e. Regression

Galton's "law of regression" refers to the widespread phenomenon already explained of a constant
swinging back to mediocrity which the breeder must
oppose with continual selection in order to maintain
the standard of any particular strain.   We have
seen that within a "pure line," regression is complete
and that in populations made up of a mixture of
pure lines it is a factor always to be reckoned with.
Regression, however, has to do with fluctuating variations and does not bring about a permanent change
of type.   It should, therefore, not be confused with
reversion.

3. Explanation of Reversion

Darwin, who did not always differentiate between
reversion and atavism, suggested that reversion was

due sometimes to the action of a more natural environment, as in the case of animals set free after having been in captivity, and sometimes to hybridization, since there seems to be a general tendency of hybridized organisms to "revert" to ancestral types.

It is now known that reversion, like atavism, is simply a case of latent characters becoming apparent according to the Mendelian principle of segregation. To quote Davenport: "There is nothing more mysterious about reversion, from the modern standpoint, than about forming a word from the proper combination of letters."

4. SOME METHODS OF IMPROVING OLD AND ESTABLISHING NEW TYPES

a. The Method of Hallet

This method, which was formulated by the English wheat-grower Hallet in 1869, has been in common use for a long time. It consists in placing the organisms to be bred in the very best possible environment and then choosing those individuals which make the best showing as the stock from which to breed further, a procedure based upon the deep-seated belief that acquired characters are inherited.

For example, in a field of wheat, plants near the edge of the field which, from lack of crowding or by reason of proximity to an extra local supply of fertilizer or any other favorable environmental factor, make a more vigorous growth than their neighbors, are selected in the hope that the gains made by them will be maintained in their offspring.

We have seen that it is very questionable whether acquired characters which are due to environmental conditions play any rôle whatever in heredity. The phenotypic character does not always indicate what the germplasm will subsequently do, and when the true genotypic constitution of the germplasm is still further masked by the temporary fluctuations caused by a modified environment, it is increasingly difficult to select wisely from the display of variants those which will produce the best ancestors for the future stock.

That this common procedure of selecting the best-appearing animal in the flock and the biggest ear of corn in the bin, has met with a large degree of success in the past is due entirely to the fact that in many instances the phenotypic character is an actual expression of the genotypic constitution. This is not always the case, however, and we cannot now fail to see that the method is blind and full of error. Its successes are due to the indirect results of chance rather than to a direct control of the factors of heredity. The great proportion of failures resulting from this procedure now find a reasonable explanation from the standpoint of Mendelism.

b. The Method of Rimpau

Contrasted with the Hallet method of augmenting acquired characters and then selecting the best display of them, is the method of Rimpau, who experimented for two decades with various grains and, finally, among other results, produced the famous Schlandstedt barley.

Rimpau's method is to sow grain under ordinary conditions with a minimum rather than a maximum amount of fertilizer and then to select individuals, neither from the rich spots nor from the edges of the field where there is little crowding, but from situations where the environmental conditions are ordinary or even unfavorable. Individuals making a good showing under such usual, or even adverse, conditions are worthy by *nature* rather than by *nurture* and are consequently most desirable as progenitors of future stock. By this method the attempt is not to keep the progeny of single individuals separate, but to mass together the best as they appear under ordinary normal environment.

This again is an indirect method of procedure, although the character of the germplasm is more nearly hit upon in this way than by Hallet's method, since the mask of temporary accessory modifications is stripped so far as possible from the somatoplasm, and the phenotype made to approximate the genotypical constitution.

c. The Method of de Vries

The method of de Vries has already been in part described in Chapter IV. It depends upon the preservation and exploitation of the mutations occurring in nature. It recognizes clearly the fact that change of type is dependent upon a germplasmal variation which is largely, if not entirely, independent of environmental factors.

Accordingly, the work of the successful breeder consists in simply taking what nature spontaneously furnishes to him rather than in attempting to force nature into producing something new. These mutations, when isolated, may become the progenitors of desirable new lines.

d. The Method of Vilmorin

This is an isolation method which has been successfully applied to the sugar-beet industry. The seeds from each plant to be tested are sown in separate beds from which upon maturity samples are taken and tested for sugar content. The plants from the bed furnishing the sample which contains the highest percentage of sugar are then used as the seed producers for the next generation. In this way by continual selection an improved strain may be maintained.

e. The Method of Johannsen

The method of isolating pure lines or homozygotes out of a mixed population has been considered in Chapter VI. As in the method of de Vries of isolating mutations, so, too, in the pure line method it is recognized that the germplasm is the source of initiatory changes and that the technique of establishing new types consists in sorting out homozygous strains of this germplasm.

The method of Johannsen is quite different from those of Hallet and Rimpau in that the ideal organization is not sought for among phenotypes, but among genotypes. It is not the somatoplasm, but the germplasm that is selected.

f. The Method of Burbank

This is a method of greatly increasing the number of variants by promiscuous hybridization and then of eliminating all except those of a desired phenotypic combination. Indirectly it depends upon the principle of the segregation of unit characters which makes possible *rearrangements* of these characters according to the laws of chance. The characters themselves remain unchanged, since nothing new is produced by hybridization except *new arrangements* of existing characters.

The spectacular success of Luther Burbank in "creating" new plant forms is due largely to his very extensive hybridizations, his skill in detecting among the varying progeny the winning phenotype and his ruthless elimination of the great majority of variations that do not quite fill his requirement.

The successful combinations must be propagated in most instances asexually by grafting, cuttings, bulbs, etc., rather than sexually through the medium of seeds, because new genotypes which will breed true are not necessarily isolated by this procedure. The consequence is that Burbank's method cannot be utilized in animal breeding to any great extent where the maintenance of a desirable strain by asexual propagation is out of the question.

It will be seen that this method, like the first two, is

fortuitous and to a certain extent unscientific in that
no one can repeat the exact conditions of the experiment and arrive at the same results.   It depends upon
the chance mixing up of a large number of possibilities
and then in not being distracted or blinded by the
good while selecting the *best*.   In the hands of a skilful
plant breeder with unlimited resources at his com-
mand it may result in much practical achievement, but
it does not particularly illuminate the path of other
breeders who wish to repeat the experiment.   It is
after all a selection of phenotypes and, therefore,
forever open to error, since phenotypes do not always
indicate what the behavior of their constituent genotypes will be in heredity.

g. The Method of Mendel

The method of Mendel, like the foregoing, depends
upon hybridization with the difference that the
desired combination is sought directly by definite
predetermined crosses, according to the expectations
of the Mendelian ratios, rather than through the
random result of fortuitous combinations.   This
method has been rendered possible by the determination of Mendel's laws of dominance, and of the independence and segregation of unit characters which
give to the experimental breeder definite expectations
and a method of procedure.

If, upon hybridization, the desired character behaves like a *recessive*, then all that is necessary to
establish a pure stock exhibiting the character in
question, is to breed two recessives together, because

recessives are always homozygous and, regardless of their ancestry, breed true.

On the other hand, if the desired character proves to be a *dominant*, then it is necessary to determine whether it is present in a duplex or a simplex condition; in other words, whether it is homozygous or heterozygous, for only homozygous organisms breed true. Establishing a strain consists, consequently, in making an organism homozygous.

The test to determine whether a dominant character is homozygous or heterozygous, that is, whether it will breed true or not, can be made by a single cross according to the procedure outlined in paragraph 8 of Chapter VII. If, upon crossing the individual to be tested with a recessive, it produces an entirely dominant progeny, then its germplasm is duplex for this character, and it will always reproduce the character in either duplex or simplex condition according to what it may be crossed with. When crossed, for instance, with another duplex dominant like itself, a pure homozygous strain of the character in question will be perpetuated.

If, on the contrary, the dominant character to be tested proves to be simplex or heterozygous, as determined by the fact that, when crossed with a recessive, 50 per cent of the progeny are recessive, then it requires more than a single generation to establish a homozygous dominant strain.

In random inbreeding of diverse strains *if the recessives are constantly eliminated as they appear*, a population is gradually obtained which is composed of an increasing number of dominants so that after only a few generations the chances are much reduced that recessives will appear, which means the practical purity of the strain.

5. The Factor Hypothesis

It has been ascertained within the last decade that some characters require more than a single determiner to bring them to expression. The idea of compound determiners for a single character may be termed the *factor hypothesis* of heredity. The converse is also true, that certain single determiners may control more than one character. For instance, the determiner for gray hair in rats also produces a lighter color on the belly.

Mendel, whose experiments led him to believe that each character depends upon only a single determiner for the reason that he worked on characters severally belonging to different parts of the plant, was apparently unaware of the existence, in certain cases at least, of compound determiners.

These compound factors may be arranged in various categories. For example, there may be, —

(1) A *complementary factor* which is added to a dissimilar factor in order that a particular character may appear;

(2) A *supplementary factor* which is added to a dissimilar factor with the result that a character is modified in some way;

(3) A *cumulative factor* which, when added to

another similar factor, affects the degree of expression that a character is given;

(4) An *inhibitory factor*, which prevents the action of some other factor, and so on.

It will be profitable to consider the factor hypothesis in some detail, since it helps to explain both reversion and the formation of new types.

a. Bateson's Sweet Peas

In the course of numerous breeding experiments Bateson obtained two strains of white sweet peas, *Lathyrus*, which, when normally self-fertilized, each bred true to the white color. When these two strains were artificially crossed, however, the progeny all had purple flowers like the wild ancestral Sicilian type of all cultivated varieties of sweet peas.

Here was apparently a typical instance of reversion, but according to the factor hypothesis the explanation is this. The character of purple color is dependent upon two independent factors which, though separately heritable, are both required to produce it. Each of these white strains of sweet peas possesses one of these factors which can produce colored flowers only when united with its complement, a proof of which appeared upon interbreeding hybrid purples from such a cross. In short, the color purple depends upon the action of two complementary factors which follow the behavior of a dihybrid. (See Chap. VII, par. 10.)

The gametic formulæ for the two strains of white sweet peas used in this experiment are Cp and cP, respectively. C stands for a color factor without which no color can appear, even though pigment for color may be present, and c is the absence of this factor, while P represents a purple pigment factor which only finds expression in the somatoplasm when

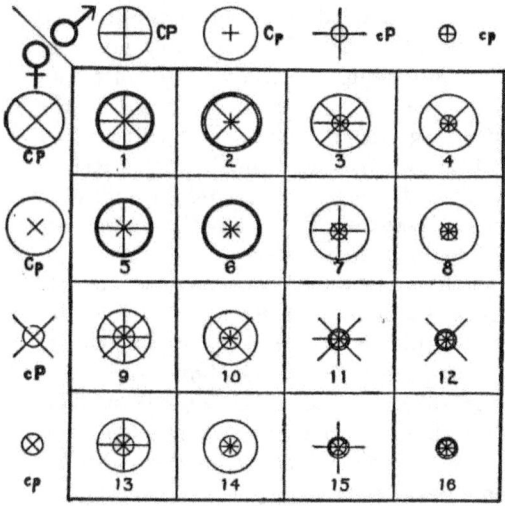

FIG. 49. — Diagram to illustrate the possible progeny from two heterozygous purple sweet peas according to data from Bateson. C, color factor (large circles) ; c, absence of C (small circles) ; P, pigment factor (large crosses) ; p, absence of P (small crosses). In the zygotes within the checkerboard squares the gametic symbols are superimposed.

taken together with the color factor C. The small letter p stands for the absence of the purple pigment factor. It will be seen that each of the white sweet peas whose formulæ are given above lack one of the two essential factors for purple color. When the

M

two are crossed, however, all the progeny are purple with the formula $CcPp$.

These hybrid sweet peas upon gametic segregation theoretically produce four kinds of gametes, CP, Cp, cP, and cp which may combine as any other dihybrid in sixteen different ways. In this case, however, these combinations group themselves into only two phenotypes, purple and white, as indicated in the accompanying diagram (Fig. 49) in which C and c are represented by large and small circles respectively, while P and p are correspondingly indicated by large and small crosses. The gametic symbols are superimposed to form the zygotes.

The theoretical expectation here shown was closely approximated in the actual results.

It may be noted in passing that the seven kinds of white sweet peas resulting from the above cross, while phenotypically alike, that is, in the zygotic symbols of Figure 49, lacking either the large circle (color) or the large cross (pigment), belong to three distinct genotypes as follows : —

		Number of Zygote in Figure 49
1	Without the pigment factor (large cross)	6 · 8 · 14
2	Without the color factor (large circle)	11 · 12 · 15
3	Without either pigment (large cross) or color (large circle)	16

Among the purple peas are the following four genotypes : —

		Number of Zygote in Figure 49
1	Duplex for both color (large circle) and pigment (large cross)	1
2	Duplex for color (large circle) but simplex for pigment (large cross)	2·5
3	Simplex for color (large circle) but duplex for pigment (large cross)	3·9
4	Simplex for both color (large circle) and pigment (large cross)	4·7·10·13

b. Castle's Agouti Guinea-pigs

An illustration of a *supplementary factor* that acts only in conjunction with some other to bring about a modification, is the *pattern factor* demonstrated by Castle in his guinea-pigs.

The wild gray, or "agouti," color of the hair of certain guinea-pigs is due to the fact that pigment is distributed along the length of each hair in a definite pattern. The tip of a single hair is black followed by a band of yellow, while most of the proximal part which is more or less concealed by overlapping hairs is a leaden color. The distribution of pigment in such a pattern gives the characteristic gray, or agouti color to the coat when taken as a whole.

Castle demonstrated the separate nature and behavior of such a pattern factor when he discovered that it is transmitted *independently of pigment*, which is necessary to bring it to expression. He showed that upon crossing a solid black guinea-pig, unquestionably possessing pigment but no "pattern," with a white albino guinea-pig having no pigment, some of the offspring "reverted" to the ancestral agouti, or "pattern" type, thus proving that the pattern must be carried in this case by the white or albino guinea-pig as a factor independent of the color which is necessary for its expression.

c. Cuénot's Spotted Mice

Another instance of the interaction of supplementary factors is seen in the spotting of piebald mice. Cuénot discovered that such spotting is due to the absence of a *uniformity factor* which if present causes color to be uniformly distributed over the entire coat.

Both of these independent factors, spotting and uniformity, are real and not imaginary, since they may be separately transmitted through albino animals in the same way as the pattern factor mentioned above, notwithstanding that in albinos both are hidden through the absence of pigment, upon the presence of which their visibility depends.

Whenever piebald or spotted animals appear in a progeny derived originally from self-colored stock, it is evidently due to the absence of such a "uniformity" factor as has just been described.

Galton's theory of "particulate inheritance" (page 121) is now satisfactorily explained as true alternative inheritance in which the mosaic appearance is caused by a Mendelian determiner, in this instance a spotting factor or, in other words, the absence of a factor for uniformity.

d. *Miss Durham's Intensified Mice*

Miss Durham, in her work with mice, has demonstrated an *intensifying factor*, the absence of which she calls a *diluting factor*. The action of the former produces, as its name implies, intensity of color, while that of the latter serves to lessen the degree of intensity in which color appears.

These factors of intensity and diluteness, it should be observed, do not in any way correspond to the duplex and simplex condition of a dominant color character, either of which would straightway appear if crossed with an albino. The factors of intensity and dilution of color are of an entirely different nature, as they have been proven to be independently transmissible through albinos where a color character could not appear because of the absence of pigment.

The following illustration of this kind of supplementary factors taken from Miss Durham's experiments will serve to make the case clear. The symbols employed are : —

B = *black* pigment which masks brown, or chocolate.

b = the absence of B, consequently *chocolate*.

I = intensity factor.

i = dilution factor or absence of intensity.

C = a complementary color factor acting with P.

P = a complementary pigment factor acting with C.

$BICP$ = black.

$BiCP$ = blue or maltese (dilute black).

$bICP$ = chocolate.

$biCP$ = silver-fawn (dilute chocolate).

The crosses which were made are represented in the table below, in which the expectation according to the Mendelian dihybrid ratios is given in parentheses after the actual results of each cross.

	BLACK (BICP)	BLUE (BiCP)	CHOCO- LATE (bICP)	SILVER- FAWN (biCP)
Black (BICP) × Silver-fawn (biCP)	9(9)	4(3)	3(3)	2(1)
Blue (BiCP) × Chocolate (bICP)	42(45)	16(15)	14(15)	8(5)
Blue (BiCP) × Silver-fawn (biCP)	0(0)	33(36)	0(0)	12(12)

It will be seen that the actual results, even when such small totals are concerned, approximate very closely the expectation and are entirely consistent.

e. Castle's Brown-eyed Yellow Guinea-pigs

Recently Castle has shown that in guinea-pigs there is an independent factor for *extension* of pigment distinct from the uniformity factor already mentioned. The absence of this extension factor ("restriction") is manifested by a lack of black or brown pigment everywhere except in the eyes and to a slight extent in the skin of the extremities, *while the distribution of yellow is wholly unaffected by it.*

That such "extension" and "restriction" factors really exist, is proven in the following way : —

When a brown (chocolate) guinea-pig is crossed with an ordinary black-eyed yellow one, the young are all black pigmented, but by cross-breeding

these hybrid young four varieties are obtained in
the next generation, viz., black, brown, black-eyed
yellow, and *brown-eyed yellow*, the latter a variety
unknown before Castle's experiment in breeding
was made.

For the sake of clearness the formation of the
brown-eyed yellow is shown below in Figure 50.

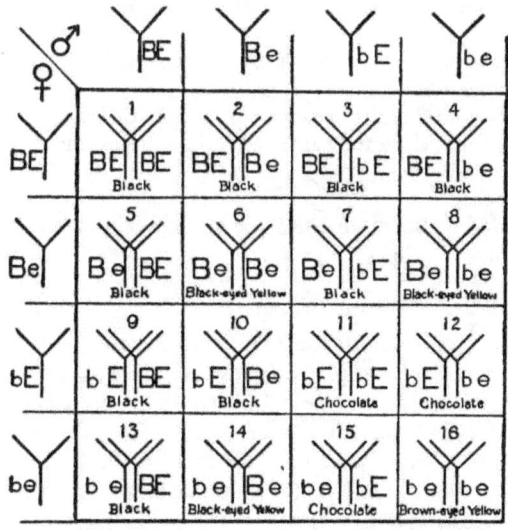

FIG. 50. — Diagram to illustrate the origin of a brown-eyed yellow guinea-
pig from two heterozygous black parents based upon Castle's experiments. The factor for yellow (*Y*) is present in every gamete and is
consequently duplex in every zygote but is hidden whenever the factor *B* is present. *B*, black pigment hiding brown or chocolate; *b*,
chocolate (absence of *B*); *E*, extension of *B* over the entire body
hiding *Y*; *e*, restriction of *B* to eyes alone thus exposing *Y* over the
entire body.

Symbols

B = *black* pigment, hiding brown or chocolate.

b = absence of B, or *chocolate*.

Y = yellow pigment, hidden by B.

E = extension of B over entire body, hiding Y.

e = restriction of B to eyes alone, thus exposing Y over the entire body.

C = complementary color factor acting with P to produce color.

P = complementary pigment factor acting with C to produce color.

(The factors C and P may be omitted for the sake of simplicity, since they are present in each instance.)

First Cross

"Extended" chocolate (bEY) × black-eyed yellow (BeY) = black $(BbEeYY)$.

Second Cross

When these cross-breds are mated with each other, they each form four kinds of gametes, BEY, BeY, bEY, and beY, which unite into sixteen theoretical genotypic possibilities, shown in Figure 50. These fall into four phenotypes, nine black (BEY), three black-eyed yellow (BeY), three chocolate (bEY), and one brown-eyed yellow (beY). The actual results in Castle's experiments gave all four kinds in close numerical agreement with this expectation. The action of extension and restriction factors is, therefore, plainly a case of Mendelian dihybridism in which two independent pairs of alternative characters are concerned.

6. RABBIT PHENOTYPES

Perhaps no better application of the factor hypothesis may be found than the case of the color of rabbits.

There are many varieties of rabbits so far as color is concerned, particularly among domesticated races. These varieties are now quite explainable by the factor hypothesis, as indicated in the table below. The sixteen kinds of rabbits there catalogued have

THE FACTOR HYPOTHESIS APPLIED TO COLORS OF RABBITS

CONSTANT FACTORS			ALTERNATIVE FACTORS					GAMETIC FORMULA	PHENOTYPIC CHARACTER WHEN CROSSED WITH THE SAME KIND OF GAMETIC COMBINATION
1	2	3	4	5	6	7	8		
Br	B	Y	C	E	I	U	A	$AUIEC$ [$YBBr$]	Gray
							a	$aUIEC$ [$YBBr$]	Black
						u	A	$AuIEC$ [$YBBr$]	Gray spotted
							a	$auIEC$ [$YBBr$]	Black spotted
					i	U	A	$AUiEC$ [$YBBr$]	Blue-gray
							a	$aUiEC$ [$YBBr$]	Blue (Maltese)
						u	A	$AuiEC$ [$YBBr$]	Blue-gray spotted
							a	$auiEC$ [$YBBr$]	Blue spotted
				e	I	U	A	$AUIeC$ [$YBBr$]	{ Yellow (with white belly and tail)
							a	$aUIeC$ [$YBBr$]	{ Sooty yellow (with yellow belly and tail)
						u	A	$AuIeC$ [$YBBr$]	Yellow spotted
							a	$auIeC$ [$YBBr$]	Sooty yellow spotted
					i	U	A	$AUieC$ [$YBBr$]	Cream
							a	$aUieC$ [$YBBr$]	Pale sooty yellow
						u	A	$AuieC$ [$YBBr$]	Cream spotted
							a	$auieC$ [$YBBr$]	Pale sooty yellow spotted

been obtained by Castle and other experimental breeders as well as many of the albino types that would double this list if c, or the factor for absence of color, should be substituted for C, the presence of color, in column 4 of the table on page 169.

Explanation of Symbols in the Foregoing Table

Br = a factor acting on C to produce *brown* pigmentation.

B = a factor acting on C to produce *black* pigmentation.

Y = a factor acting on C to produce *yellow* pigmentation.

The three factors, Y, B, Br, are present in every rabbit gamete and up to date have not been separable as independent unit characters, although they have been separated out in guinea-pigs and mice. There are no brown rabbits, because black always goes linked with brown covering the brown factor. Yellow rabbits result, as explained below, through the action of factor e.

C = a common *color* factor necessary for the production of any pigment. It was discovered in 1903 by Cuénot.

c = the absence of C which results in albinos, regardless of whatever pigment factors may be present. By changing C to c, sixteen kinds of albinos would be added to this catalogue, an addition of one phenotype and sixteen genotypes, all looking alike but breeding differently.

E = a factor governing the *extension* of black and brown pigment, *but not of yellow*.

e = the absence of extension or *restriction* of black and brown pigment to the eyes and the skin of the extremities only, while yellow remains extended and visible. Demonstrated by Castle in 1909.

I = an *intensity* factor which determines the degree of pigmentation. It can be transmitted independently of C through an albino. Discovered by Bateson and Durham in 1906.

i = the absence of intensity or *dilution*. Dilute black = blue. Dilute yellow = cream. Dilute gray = blue-gray.

U = a factor for uniformity of pigmentation or "self-color" discovered by Cuénot in 1904.

u = the absence of uniformity which results in *spotting with white*.

A = a pattern factor for agouti, or wild gray color, which causes the brown and black pigments to be excluded from certain portions of each hair, resulting in the gray coat. When present in the rabbit, it is also associated with white or lighter color on the under surfaces of the tail and belly. It was demonstrated by Castle in 1907.

a = the absence of the agouti or pattern factor.

7. THE KINDS OF GRAY RABBITS

Each of the apparent kinds of gray rabbits indicated in the foregoing table may be made up of various genotypes. For instance, there are thirty-two different genotypes, each of which is phenotypically a gray rabbit. The zygotic formula for each of these thirty-two possibilities is displayed in the next table, and it will be seen that these range all the way from rabbits homozygous in all their variable characters (No. 1) to those homozygous in none (No. 32).

The progeny of these various types of gray rabbits when inbred will consequently vary from the pure

The Kinds of Gray Rabbits (Color only)

#	Genotype Zygotic Formula	Number of Heterozygotic Factors	Gray	Black	Gray spotted	Black spotted	Blue gray	Blue	Blue gray spotted	Blue spotted	Yellow	Sooty	Yellow spotted	Sooty spotted	Cream	Cream spotted	Pale sooty	Pale sooty spotted	White
1	AAUUIIEECC [YBBr][YBBr]	None	×																
2	AAUUIIEECc [YBBr][YBBr]	One	×																×
3	AAUUIIEeCC [YBBr][YBBr]	One	×								×								
4	AAUUIiEECC [YBBr][YBBr]	One	×				×												
5	AAUuIIEECC [YBBr][YBBr]	One	×		×														
6	AaUUIIEECC [YBBr][YBBr]	One	×	×															
7	AAUUIIEeCc [YBBr][YBBr]	Two	×								×								×
8	AAUUIiEECc [YBBr][YBBr]	Two	×				×												×
9	AAUuIIEECc [YBBr][YBBr]	Two	×		×														×
10	AaUUIIEECc [YBBr][YBBr]	Two	×	×															×
11	AAUUIiEeCC [YBBr][YBBr]	Two	×				×				×				×				
12	AAUuIIEeCC [YBBr][YBBr]	Two	×		×						×		×						
13	AaUUIIEeCC [YBBr][YBBr]	Two	×	×							×	×							
14	AAUuIiEECC [YBBr][YBBr]	Two	×		×		×		×										
15	AaUUIiEECC [YBBr][YBBr]	Two	×	×			×	×											
16	AaUuIIEECC [YBBr][YBBr]	Two	×	×	×	×													
17	AaUuIiEECC [YBBr][YBBr]	Three	×	×	×	×	×	×	×	×									
18	AaUuIIEeCC [YBBr][YBBr]	Three	×	×	×	×					×	×	×	×					
19	AaUuIIEECc [YBBr][YBBr]	Three	×	×	×	×													×
20	AaUUIiEeCC [YBBr][YBBr]	Three	×	×			×	×			×	×			×		×		
21	AaUUIiEECc [YBBr][YBBr]	Three	×	×			×	×											×
22	AaUUIIEeCc [YBBr][YBBr]	Three	×	×							×	×							×
23	AAUuIiEeCC [YBBr][YBBr]	Three	×		×		×		×		×		×		×	×			
24	AAUuIIEeCc [YBBr][YBBr]	Three	×		×						×		×						×
25	AAUuIiEECc [YBBr][YBBr]	Three	×		×		×		×										×
26	AAUUIiEeCc [YBBr][YBBr]	Three	×				×				×				×				×
27	AAUuIiEeCc [YBBr][YBBr]	Four	×		×		×		×		×		×		×	×			×
28	AaUUIiEeCc [YBBr][YBBr]	Four	×	×			×	×			×	×			×		×		×
29	AaUuIIEeCc [YBBr][YBBr]	Four	×	×	×	×					×	×	×	×					×
30	AaUuIiEECc [YBBr][YBBr]	Four	×	×	×	×	×	×	×	×									×
31	AaUuIiEeCC [YBBr][YBBr]	Four	×	×	×	×	×	×	×	×	×	×	×	×	×	×	×	×	
32	AaUuIiEeCc [YBBr][YBBr]	Five	×	×	×	×	×	×	×	×	×	×	×	×	×	×	×	×	×

gray, as in No. 1, to a gray from which sixteen possible types of young may be expected as in No. 32.

Up to the time when Castle's paper upon the factor hypothesis [1] was published in 1909, nine genotypic kinds of gray rabbits had been obtained in his experiments, whose genotypic formulæ correspond to the following numbers in the list: 1, 3, 6, 10, 13, 20, 22, 28, 29.

8. Conclusion

That a relatively small number of factors may produce an extensive array of combinations is evident from this data.

The analysis of germplasm by the factor hypothesis is now being generally applied by geneticists to the particular organisms with which they are concerned. It has been carried out notably in detail by both Bateson and Davenport for poultry and by Baur for the snapdragon, *Antirrhinum*.

Finally, the elucidation of the factor hypothesis makes any further explanation of reversion superfluous. It is now easy to see how a particular character may remain latent for generations and at last come to expression only when the missing factor necessary to its activity is supplied by some cross.

It is also clear how hybridization, in which many characters are concerned, is bound to furnish far more new combinations than would, at first thought, be expected.

---

[1] "Studies of Inheritance in Rabbits." Carnegie Institution Publications, No. 114, 1909. W. E. Castle in collaboration with Walter, Mullenix and Cobb.

CHAPTER IX

BLENDING INHERITANCE

1. Relative Value of Dominance and Segregation

Of the three fundamental principles which underlie "Mendel's law," namely, segregation, independence of unit characters, and dominance, the principle of dominance has been found to hold true in a surprising number of cases and in relation to very diverse organisms, notwithstanding the fact that the time spent in the investigation of dominance, as that term is now understood, has been comparatively short. Doubtless future experimentation will demonstrate the existence of dominance to a far greater extent than has at present been discovered.

Its universal application is by no means assured, however, since the mathematical precision with which it works, that following its discovery in 1900 has so captivated the biological world, is beginning to give way in the face of many exceptions which have been steadily accumulating.

Even Mendel himself noted certain exceptions to the law of dominance, and his followers have pointed out with increasing emphasis that it is subject to many modifications. It is now understood, indeed, that *segregation, not dominance,* is the most essential factor in the Mendelian scheme.

2. Imperfect Dominance.

It frequently occurs that dominance is so imperfect that a heterozygous, or simplex, dominant may be distinguished at once by simple inspection from a homozygous, or duplex, dominant, whereas the test of crossing with a recessive is necessary whenever dominance is complete, as has been previously explained. The single dose of the determiner in such a case has plainly, then, less phenotypic effect than a double dose.

There are many cases of imperfect dominance among flowering plants. Correns has shown that when plants of a white-flowering race of the "four-o'clock," *Mirabilis jalapa,* are crossed with those of a red-flowering race, all the offspring in the first filial generation, unlike either parent, exhibit rose-colored flowers. When, however, these rose-colored flowers are crossed with each other, they produce red, rose, and white in the Mendelian ratio of 1 : 2 : 1; that is, three colored to one white. The red-flowering race thus proves to be homozygous and the rose-flowering race heterozygous. Here color dominates the absence of color, or white, but the *degree* of the color depends upon whether the dose of pigment is duplex or simplex.

A classic illustration of imperfect dominance among animals is the "blue Andalusian fowl," the hereditary behavior of which is illustrated below (Fig. 51). It will be seen that when two blue Andalusian fowls, characterized by a mottled plumage, are bred together, they produce three kinds of offspring in the ratio of 1 : 2 : 1. Twenty-five per cent are clear black, 50 per cent are blue Andalusian, and 25 per cent are white "splashed" with black. Both the black and the splashed white fowls from this cross prove, upon further breeding, to be homozygous, while the blue Andalusian itself is heterozygous and can, therefore,

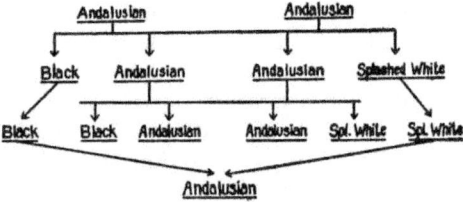

Fig. 51. — The heredity of the blue Andalusian fowl, an illustration of "imperfect dominance."

never be made to breed true. In order to produce 100 per cent of blue Andalusian chicks, it is necessary simply to cross a splashed white with a black Andalusian.

There is nothing in this case to indicate whether the black or the splashed white should be regarded as the homozygous dominant, since dominance is imperfect. In either case the heterozygous blue Andalusian is at once evident in the first filial generation without further crossing.

A similar case of imperfect dominance is furnished by the roan color of cattle which results when red and white are crossed. If two roans are mated, they produce red, roan, and white offspring in the proportion of 1 : 2 : 1, thus showing that roan is a heterozygous character in which the dominance of red is imperfect.

Even in cases of apparently perfect dominance it is sometimes possible by close, inspection to detect differences between a pure dominant (DD), Figure 43, and a heterozygous dominant (DR) when a superficial examination is not sufficient to distinguish them.

For instance, in the cross between smooth and wrinkled peas, a microscopic examination of the starch-grains in the cotyledons of · the hybrid peas shows that they are of two kinds. Darbishire calls attention to the fact that, in the power of absorption, hybrid smooth peas (DR) are intermediate between their pure dominant smooth (DD) and pure recessive wrinkled (RR) parents.

3. DELAYED DOMINANCE

A character which is really dominant is sometimes so late in manifesting itself in the individual growth of the offspring that it may properly be termed a *delayed dominant*.

Dark-haired individuals often do not acquire their definitive hair color until adult life, and it is common knowledge that the eyes of an infant for a considerable period provoke no little speculation among adoring relatives as to " whose eyes " they are.

According to Davenport, when a white Leghorn fowl is crossed with a black Leghorn, white being dominant in this case, chicks are produced that are

N

white with black flecks in their plumage. These black
flecks, however, disappear at the time of the first
molt. The complete dominance of white is, therefore, simply delayed.

4. "Reversed" Dominance

In certain instances there seems to be a reversal
of dominance, as may be illustrated by Lang's results
with snails (*Helix*). He has proven in his experiments
that red snails are generally dominant over yellow
snails, although in certain cases there is apparently
an exception to the rule, for snails with yellow shells
dominate those with red shells.

Davenport also has shown that although extra
toes are usually dominant over the normal number
in poultry, yet, in something like 20 per cent of the
cases, the normal number is dominant.

To speak of these cases as instances of "reversed
dominance," is open to serious objection, since such
an explanation does not agree with the generally
accepted "presence and absence" idea of heritable
characters. It is difficult to see how the presence of
a certain determiner can dominate in a part of the
offspring of any cross and the absence of the same
determiner be able to dominate the remainder.

It is perhaps nearer the truth to conceive that in
cases of apparent "reversal" of dominance there is
an insufficient amount of a particular determiner
available to bring the character concerned into
expression. In other words, although a dominant
character may be present in two cases, yet in one

it fails, for some reason, to become effective. This interpretation agrees with the facts brought out by subsequent breeding in cases of this sort.

It sometimes occurs that a character which is dominant in one species may be recessive in another. Horns are dominant in sheep, but recessive in cattle. White color is recessive in rodents and sheep, but dominant in most poultry and in pigs.

5. POTENCY

Davenport seeks to explain modifications in typical dominance as variations in the *potency* of determiners. He defines potency as follows: "The potency of a character may be defined as the capacity of its germinal determiner to complete its entire ontogeny."

That is, if the potency of a determiner, for some reason, is insufficient, there may be either an incomplete or delayed manifestation of the character in question, or it may fail entirely to develop.

The variations of potency may be grouped into three general categories according to the degree of their manifestation; namely, total potency, partial potency, and failure of potency.

A further word of explanation for each of these three kinds of potency seems desirable at this point.

a. Total Potency

This is complete Mendelian dominance in which even the heterozygotes produced by a simplex dose of a character are indistinguishable phenotypically, that is, by inspection, from the homozygotes produced by a duplex dose of the same character. It is as if a single bottle of black ink poured into a jar of water was just as effective as *two* bottles of ink, in forming an opaque fluid.

b. *Partial Potency*

Partial potency covers all cases of *incomplete dominance*, such as those of the four-o'clock (*Mirabilis*) and blue Andalusian fowls, where a simplex dose of a determiner does not produce the same visible effect as a double dose.

The dominant prickly Jamestown weed (*Datura*), when crossed with a recessive glabrous variety of the same plant, produces cross-breds in the first generation which show only a few prickles (Bateson) (Baur), following the law of partial potency.

Banded and uniformly colored snails also, when crossed together, produce snails with shells showing only a pale banding (Lang).

Numerous further instances of incomplete dominance could be cited.

c. *Failure of Potency*

If for any reason a determiner fails to accomplish its possibilities in whole or in part, then the character in question may never become evident, and the result, so far as appearances go, is the same as if it was a recessive lacking the determiner entirely.

That the failure of potency is not identical with the absence of a determiner can usually be demonstrated by further breeding, because dominants failing in potency, which are either of the formula DD or DR, may, if bred *inter se*, give a various progeny among which the dominant character D is likely to again become manifest, while recessives, of the formula RR, on the contrary, will always give offspring which all agree in the entire absence of the character in question.

Davenport cites an extreme case of failure of potency in one of two rumpless cocks from the same blood. The character of rumplessness is due to an *inhibitor* of tail development. That these two cocks both possessed this character was demonstrated by the entire absence of any tail in either case. The inhibiting determiner for tail growth was so weak in cock No. 117, however, that, to quote Davenport's exact words : "In the heterozygote the development of the tail is not interfered with at all, and even in extracted dominants it interfered little with tail development, so that it makes itself felt only in the reduced size of the uropygium and in-bent or shortened back. But in No. 116 the inhibiting determiner is strong. It develops fully in about 47 per cent of all the heterozygotes and in extracted dominants may produce a family in *all* of which the tail's development is inhibited."

Here were two birds of the same blood, phenotypically alike and presumably genotypically alike, which because of an individual difference in the potency of the determiner for rumplessness produced quite different results in their offspring although bred to precisely the same array of hens.

6. Blending Inheritance

In the instances of imperfect dominance given above, where the progeny of unlike parents present an intermediate condition, it is found that, upon cross-breeding these offspring, segregation into the grandparental types occurs just as truly as in instances of complete dominance.

In poultry, for example, when Cochins, which are "booted," and Leghorns, which are clean-shanked, are crossed, booting of an intermediate grade of four results, on a scale in which ten represents complete booting, and zero no booting or clean shank (Davenport). The character of booting and its alternative absence, however, segregate out in true Mendelian fashion when these hybrids are subsequently crossed together. It is evident that dominance plays only a secondary rôle in such cases, and that the all-important factor is segregation.

Are there, then, any cases where true fusion of hereditary parental traits occurs, in other words, where *segregation in the second filial generation does not appear?* Does the "melting-pot of cross-breeding" ever "melt" the characters thrown into it?

It was formerly believed that diverse parents generally produce intermediate offspring, and that this intermediate condition continues without any segregation at all in the form of "blending inheritance," but within the last decade apparent cases of blending inheritance have been thrown out of court one after the other by the Mendelians. Bateson, in an inaugural address at Cambridge University in 1908, stated that what was once believed to be the rule has now become the exception. He goes on to say: "One clear exception I may mention. Castle finds that in a cross between the long-eared lop rabbit and a short-eared breed, ears of intermediate length are produced; and that these intermediates breed approximately true."

Let us examine this "one clear exception" a little more closely.

7. THE CASE OF RABBIT EARS

As a typical example of blending inheritance in rabbit ears may be cited the following case: —

A female Belgian hare with an ear-length of 118 mm. was crossed with a male lop-eared rabbit with an ear-length of 210 mm. The average of these ear-lengths is 164 mm. Five offspring from this pair had ear-lengths, when adult, approximating this average as follows: 170, 170, 166, 156, 170, of which two were females and three were males. When from this litter one of the females measuring 170 mm. in ear-length was subsequently crossed with her brother having an ear-length of 166 mm., two litters were produced in which the individuals when adult attained ear-lengths of 170, 166, 168, 160, 172, and 168 mm. These results are represented diagrammatically in Figure 52.

This illustration is typical of many other breeding experiments made by the same investigators[1]

[1] Castle, in collaboration with Walter, Mullenix and Cobb. "Studies of Inheritance in Rabbits." Carnegie Institution Publications, Washington, No. 114, 1909.

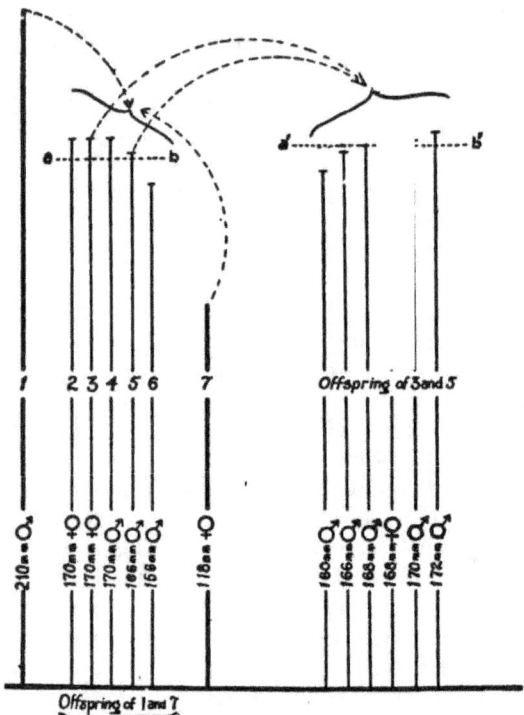

Fig. 52. — A case of three generations of ear-length in rabbits. *a–b*, average ear-length of the first filial generation (F₁). *a'–b'*, average ear-length of the F₂ generation derived from 1 and 7. Data from Castle, in collaboration with Walter, Mullenix and Cobb.

upon the ear-length of rabbits which included 70 different litters of rabbits containing 341 individuals. In none of these experiments could the blend in the second filial generation be called perfect, but it may at least be said that evidence of segregation, that is, a return to one or the other of the parental types, was much less apparent than evidence of blending.

Furthermore, crosses were made in which lop ears of various fractional lengths were obtained as desired, including $\frac{1}{8}$, $\frac{1}{4}$, $\frac{3}{8}$, $\frac{1}{2}$, $\frac{5}{8}$, $\frac{3}{4}$, and $\frac{7}{8}$ lengths. Not one of these fractional lengths apparently segregated in subsequent generations after the Mendelian fashion, but all bred approximately true.

Moreover, ears of one half lop length, for instance, were obtained in three ways: first, by crossing full-length lops with short-eared rabbits as indicated in the first cross of the case cited above; second, by crossing one half lop lengths together, demonstrated by the second cross in the illustrative case given, and third, by mating $\frac{1}{4}$ and $\frac{3}{4}$ lop lengths. Theoretically, $\frac{1}{8}$ and $\frac{7}{8}$ as well as $\frac{3}{8}$ and $\frac{5}{8}$ lop lengths would also produce $\frac{1}{2}$ lop lengths, for in all of the crosses that were made the length of ear behaved in a blending fashion.

These results were based, not upon a single measurement of each specimen, which might be open to considerable error, but upon daily measurements from the time the rabbits were two weeks old until their ears ceased to grow at about twenty weeks. The growth curves drawn from these daily measurements showed continually an intermediate or blending condition in progeny derived from diverse parents.

A Mendelian explanation of this apparently exceptional case of blending inheritance has been suggested by Lang based upon the result of Nilsson-Ehle's discoveries while breeding wheats at the Agricultural Experiment Station of Svalöf in Sweden.

8. The Nilsson-Ehle Discovery

Nilsson-Ehle found in breeding together different strains of wheat that a certain wheat with brown chaff crossed with a white-chaffed strain yielded only brown-chaffed wheat in the first generation. These heterozygous or hybrid brown-chaffed wheats when crossed with each other produced, not the expected proportion of three brown to one white, but *fifteen brown to one white*. This was not explainable as the chance result of a single cross, but was the conclusion drawn from fifteen different crosses all of the same strains that yielded a total progeny of 1410 brown-chaffed to 94 white-chaffed plants, which happens to be exactly the proportion of fifteen to one.

In other experiments it was discovered that although dominant red-kerneled strains of wheat crossed with white-kerneled varieties usually gave the three-to-one proportion upon segregation in the second filial generation, yet *one particular strain* of red-kerneled Swedish wheat in the second generation gave approximately sixty-three red to one white-kerneled strain.

The explanation of these two unexpected results is this. In the case of brown-chaffed wheat there are two independent determiners for the character of brown color, and these simply follow the Mendelian laws for a dihybrid, while in the case of the red-kerneled wheat there are three independent determiners for the character of red color, each of which is able to give red color to the wheat. Taken together, these three determiners behave cumulatively, following the law of a trihybrid.

For example, if a brown-chaffed wheat with the formula BB', in which B and B' each represent a brown-chaffed factor, is crossed with a white-chaffed wheat of the formula bb', in which b and b' each represent the absence of B and B' respectively, then all the progeny of this cross will be brown-chaffed, having the zygotic formula $BB'bb'$. When upon maturation the gametes form out of the germ-cells from such hybrids, the following four combinations are possible, and no others:

	BB'	Bb'	bB'	bb'
BB'	BB' BB' ④	Bb' BB' ③	bB' BB' ③	bb' BB' ②
Bb'	BB' Bb' ③	Bb' Bb' ②	bB' Bb' ②	bb' Bb' ①
bB'	BB' bB' ③	Bb' bB' ②	bB' bB' ②	bb' bB' ①
bb'	BB' bb' ②	Bb' bb' ①	bB' bb' ①	bb' bb' ⓪

Fig. 53. — Diagram of the possible combinations in the F_2 generation of brown-chaffed wheat according to experiments of Nilsson-Ehle. B and B' are cumulative factors for the brown-chaff character. b and b' denote the absence of B and B' respectively.

BB', Bb', bB', bb'. These represent, therefore, the possible gametes present in each sex of the first filial generation, and upon intercrossing they can combine into sixteen possible zygotes to form the second filial generation, as shown in Figure 53.

The numbers in the squares indicate how many times a brown determiner is present in each zygote. It will be seen that only one out of the sixteen possibilities lacks a brown-chaff factor, and this one will

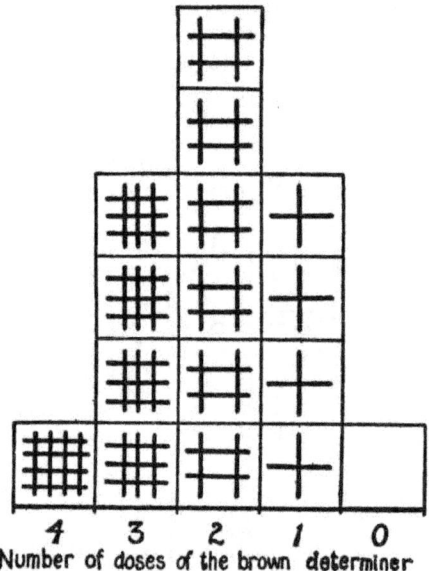

4   3   2   1   0
Number of doses of the brown determiner

FIG. 54. — The distribution of the sixteen possibilities resulting when two similar determiners (brown-chaff) act together as a dihybrid.

consequently produce only white chaff, while the remaining fifteen possibilities, each of which has at least a single determiner for brown, will all yield brown chaff.

The brown-chaff factor, moreover, is present in

varying doses among these fifteen possibilities, as indicated by the numbers in the squares. It is evident, therefore, that several shades of brown will be rep-

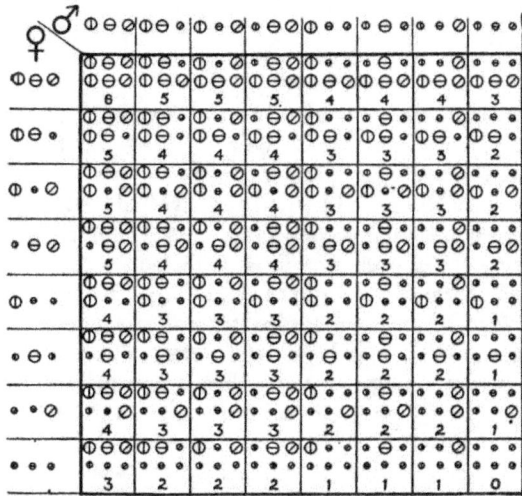

Fig. 55. — Diagram to illustrate Nilsson-Ehle's case of trihybrid red wheat. The large screwheads each represent a single determiner for the red character. The small screwheads symbolize the absence of the red character, or white. The number in each square indicates how many doses of the "red" determiner is present. For further explanation see text.

resented depending upon the number of doses of the brown determiner in each instance.

Figure 54 shows how these different shades of brown arrange themselves in the manner of a frequency polygon of fluctuating variation with the greatest number in the halfway class and the least

numbers at the two extremes. In this instance
six out of sixteen individuals of the second generation theoretically present a perfect "blend" between the original brown- and white-chaffed grandparents, although complete segregation has actually
occurred.

The same explanation holds true as displayed in
Figure 55 for the trihybrid case of red- and
white-kerneled wheats in which only one white-
kerneled to sixty-three red-kerneled individuals appear in the second filial generation. The number
of red determiners in each zygote is indicated by the
figure at the bottom of each square. The large screw-
head symbols with vertical, horizontal and diagonal
slots each represent an independent determiner for
red kernel, while the small screw heads symbolize
the absence of each of these determiners, or white
kernel. When the pure strain of red-kerneled wheat
is crossed with a pure strain of white-kerneled wheat,
the first generation is all a heterozygous red of a

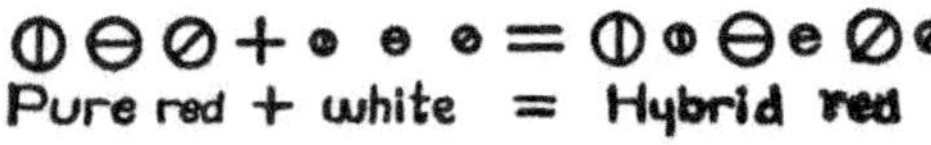

Pure red + white = Hybrid red

Fig. 56. — The result of crossing white wheat with trihybrid red wheat.

somewhat lighter shade than the original pure red
strain.

When plants of this heterozygous sort are crossed
together, they yield plants producing red-kerneled
and white-kerneled wheats in the proportion of sixty-
three to one. The sixty-three kinds of red wheats are

of varying degrees of redness and may be classified after the manner of fluctuating variations with the greatest number of kinds at the intermediate degree between pure red and pure white. (See Figure 57.)

In order to test whether the sixty-four kinds of wheats produced in the second filial generation, as theoretically displayed in Figure 55, really contain separable, though indistinguishable, determiners for red-kernel, Nilsson-Ehle produced families of the third filial generation by self-crossing plants of the second generation. It was to be expected that, if these hybrid wheats of the second generation carried one, two, three, or more determiners for a red kernel as the theoretical tables in Figures 55 and 57 demand, their progeny would be distributed with reference to the number of red- and white-kerneled individuals, in the following ratios : —

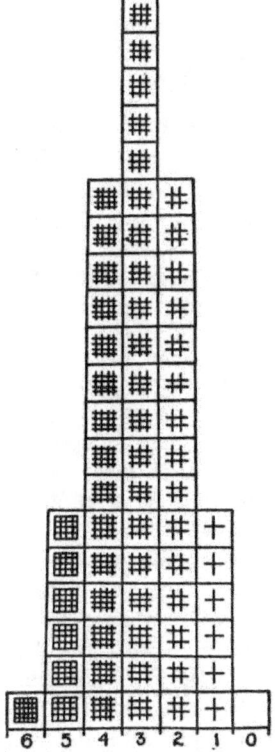

Fig. 57. — The distribution of the sixty-four possibilities in the F_2 generation when three similar determiners act together as a trihybrid.

3 red to 1 white when 1 heterozygous determiner for red
    is present.

15 red to 1 white when 2   ) heterozygous deter-
63 red to 1 white when 3   }   miners for red are
All red to no white when 4 or more)   present.

Among seventy-eight sample families of the third
generation inbred to test this theoretical conclusion,
the actual results were : —

8 families giving the ratio of 3 red to 1 white.
15 families giving the ratio of 15 red to 1 white.
5 families giving the ratio of 63 red to 1 white.
50 families giving the ratio of all red to no white.

It has been actually demonstrated therefore, in
the case of this particular strain of wheat: (1) that the
factors producing red kernel are several in number;
(2) that they act independently of each other in
heredity; (3) that these several independent factors
segregate; and (4) that any one red factor acting
alone produces a "red" result.

The Nilsson-Ehle principle of cumulative determiners has been confirmed in America by East in a masterly series of breeding experiments upon maize.

In connection with the Nilsson-Ehle principle, it
will be seen that the possible number of intergrades
between the two extremes increases rapidly as the
number of duplicate determiners increases. Thus
with six duplicate determiners for the same character
present, the ratio of possible dominants to recessives
in the second filial generation would be 4095 to 1.
The reappearance of this single recessive among 4095

dominants would be extremely unlikely, and it might easily be mistaken for a mutation or a freak. Apparent blends of all intermediate degrees, however, would be sure to appear. Yet these are not blends in the "melting-pot" sense at all, but strictly cases of Mendelian dominance and segregation.

9. The Application of the Nilsson-Ehle Explanation to the Case of Rabbit Ear-length

The so-called blending rabbit ears, along with other similar cases, can now be made to fall into line, as pointed out by Lang, with the Mendelian law of segregation.

If we assume that the long ear of the lop rabbit has only three independent but equal determiners for excess length, the case becomes one of Mendelian trihybridism with cumulative factors, which works out like Nilsson-Ehle's red-kerneled wheat in the following manner: —

In general the average for full lop ear-length may be placed at 220 mm. and for the ordinary short-eared rabbit [1] at 100 mm. The difference, or *the excess length of the lop ear*, is 120 mm., which, according to the trihybrid formula, corresponds to the six doses of the character symbolized in the upper left-hand square in Figure 55 by six large screw heads, three

---

[1] Not the Belgian hare, as cited in the illustration given in Figure 52. The Belgian hare has typically a somewhat longer ear than the ordinary short-eared rabbit.   ·

o ·

coming from each parent respectively. If all of these independent determiners are equal as regards excess ear-length, each factor would represent an excess of 20 mm. above the normal ear-length found in short-eared rabbits, that is, —

$$\frac{220 \text{ mm.} - 100 \text{ mm.}}{6} = 20 \text{ mm.}$$

When according to this computation a lop (20 mm. $\times$ 6 + 100 mm. = 220 mm.) and a pure short-eared rabbit (20 mm. $\times$ 0 + 100 mm. = 100 mm.) are crossed, if imperfect dominance occurs, which is a very common phenomenon, it is true that the offspring might present a "blended" appearance. If now these cross-breds of the first generation prove to be trihybrids with respect to excess ear-length, there would be sixty-four possibilities in their progeny segregating out just as in the red-kerneled wheat.

These possibilities would be arranged in the following frequencies: —

Number of Excess Ear-length Determiners	Number of Cases occurring out of 64	Total Length in Millimeters of Ears resulting
6	1	220
5	6	200
4	15	180
3	20	160
2	15	140
1	6	120
0	1	100

Since the average litter among rabbits is about five, the chances that these five rabbits will breed true to their hybrid parents and form a perfect blend between their grandparents is 20 out of 64, while the chance of their being like either grandparent is only one out of 64.

It should be noted further that 50 out of 64, or 77 per cent, of these hybrids of the second filial generation would have an ear-length between 140 and 180, thus approximating a "blend" closely enough to be so classified upon a casual inspection.

Moreover, if it should be found that excessive ear-length in rabbits is due to more than three duplicate determiners, the possibilities of getting anything but an apparent blend would be much decreased.

The fact, furthermore, that the fractional ear-lengths of the hybrid rabbits in Castle's experiments bred approximately true in the second and subsequent filial generations, may also be explained by the Nilsson-Ehle hypothesis.

For example, half lop lengths, according to this explanation, are those with three doses of the determiner for excess ear-length. It follows that the progeny of two rabbits each carrying three doses of a determiner will likewise, after the reduction during the maturation of the germ-cells, have three doses of the determiner $\left(\dfrac{3+3}{2} = 3\right)$.

It would be interesting to breed rabbits having ears of one eighth lop length in which, according to the foregoing hypothesis, there would presumably be present only a single determiner for excess ear-length, with ordinary short-eared rabbits having no excess ear-length, in order to see if the expected Mendelian three-to-one proportion for a monohybrid would appear in the progeny.

10. Human Skin Color

Finally, although accurate published data are wanting, it is probably true that skin color in all kinds of hybrids resulting from crosses between negroes and whites is not a case of blending inheritance, as commonly supposed, but rather of true Mendelian segregation. In fact, there is frequently visible evidence that segregation does occur, as shown by many authentic instances where the offspring of diversely colored parents produce children with skin color of different shades.

If human families included hundreds of offspring in a single generation instead of the usual number, the problem of skin color in man could doubtless be quickly solved since ratios could then be obtained large enough to reveal the underlying laws of inheritance.

CHAPTER X

THE DETERMINATION OF SEX

1. Speculations, Ancient and Modern

FROM the earliest times the desirability of controlling the sex of an unborn child, in particular instances at least, has seemed very great. Likewise the wish to be able to predetermine sex among domesticated animals has made breeders quick to grasp at every clue that promised success.

There has been no want of speculations concerning the determination of sex. J. Arthur Thomson, who with Professor Geddes wrote "The Evolution of Sex" in 1889, says: "The number of speculations as to the nature of sex has been well-nigh doubled since Dreylincourt, in the eighteenth century, brought together 262 'groundless hypotheses' and since Blumenbach caustically remarked that nothing was more certain than that Dreylincourt's own theory formed the 263rd. Subsequent investigators have long ago added Blumenbach's theory of 'Bildungstrieb' or formative impulse, to the list." It may be added in passing that the hypothesis of the determinative action of external factors upon developing germ-cells which Geddes and Thomson elaborated in the book just referred to, has, in its turn, according to most biologists, joined the long roll.

197

Hippocrates thought that sex of the offspring depends upon the relative "vigor" of the parents, while Sadler (1830) concluded that the relative ages of the two parents is the determining factor. Other writers, on the contrary, have thought that the age of the mother at the time of childbirth determines the sex of the offspring, and Thury (1863), in the days before the facts of maturation were known, ascribed the determinative factor to the relative degree of "ripeness" of the egg when fertilized. It was once assumed also that the right ovary or the right testicle is the seat of one sex and the left ovary or left testicle of the other. Galen, who did the biological thinking for several centuries of mankind, asserted that the right side of the body, "being warmer" than the left, consequently produces males. Schenk cites a most amazing bit of folk-lore to the effect that: "In Servia if a man has a stye on his eyelid he comes to the conclusion that his aunt is pregnant. If the stye is on the upper eyelid, the child will be a male; if on the lower, a female."

Modern theories of sex determination, like the earlier speculations, may be resolved into two groups, namely, those which depend upon controllable external or environmental factors such as food, climate, chemical dosage and will power, and those which depend upon internal factors at present beyond control.

2. THE NUTRITION THEORY

Of external factors which may exert a moulding influence upon the sex of the offspring, *nutrition* is possibly the most potent. This factor may be conceived to act either upon the parent previous to the maturation of the germ-cells, upon the germ-cells themselves, or upon some susceptible embryonic stage of the life cycle subsequent to that of the fertilized egg.

It has been suggested that since the egg is characterized by possibly a more advanced metabolic condition than the sperm due to the presence of the nutritive yolk, consequently the more yolk or nutrition there is, the more femaleness will characterize the egg. In other words, femaleness is a nutritive condition associated in the egg with the presence of yolk.

A generation ago Professor Schenk of Vienna, by controlling the nitrogenous diet of certain royal prospective mothers, gained a soothsayer's reputation as a prophet of sex which was based upon several correct predictions.

Of course, any prediction of sex is bound to turn out correct in 50 per cent of the cases, regardless of what it is based upon, since in man the two sexes are approximately equal in numbers. Adherents of all sorts of theories, therefore, have always been able to produce considerable "evidence" to substantiate their speculations, however crude the latter have been.

Statisticians have pointed out that in times of unusual hardship, like famine or war, when the amount of available nutrition for pregnant mothers is presumably reduced, there seems to be a preponderance of males born.

A series of nutrition experiments upon frogs performed by Born ('81), Pflüger ('82), and Yung ('85) showed that the percentage of female offspring, which normally is slightly over fifty, could be changed to over 90 per cent by regulating the food supplied to the mother before the egg-laying period. Cuénot and King, however, working independently, repeated these experiments with great care, taking into account all the eggs that were laid and not simply the ones that developed, and both obtained negative results. They concluded, therefore, that the high percentage of female tadpoles appearing in the initial experiments was due to a greater mortality among the males and not to the transformation of possible males into females.

There seems to be no doubt that nutrition may affect the percentage of those which reach maturity. If one sex requires a greater amount of nutrition than the other to carry out successfully the more strenuous metabolic changes in its life-cycle, then unequal percentages between the sexes of the survivors resulting from modified nutrition do not in any way help to solve the problem of determining the sex of the individual. In other words, the elimination of one sex through modified nutrition does not "determine" the other sex.

3. The Statistical Study of Sex

From statistical sources it has been ascertained that ordinarily there is produced a practical equality in the numbers of the two sexes.

Oesterleben in Europe summarized the data for nearly sixty million human births and found that an average of 106 males are born to every 100 females.

According to various authorities, the relative number of males per 100 females is given for horses as 99, for cattle 94, and poultry 95, while in pigs, rabbits, pigeons, and greyhounds the corresponding number of males is slightly over 100.

This practical equality of the sexes in all sorts of natural environments indicates the improbability of the assumption that external conditions determine sex.

4. Monochorial Twins

There are two kinds of twins, namely, ordinary twins, which come from two separately fertilized eggs each inclosed in its own chorion, and "identical twins," that have their origin in one egg which is inclosed in one chorion. Of the former, something like 30 per cent in man are reported as being of two sexes, thus showing that it is neither nutrition nor environment which determines sex. Usually when twins are of the same sex, they exhibit as great a range of difference in mental and physical traits as do ordinary children of the same fraternity born at different times, but occasionally "identical twins" are born, and such monochorial twins are *always of the same sex*. This is evidence that sex, like other somatic characters, is determined in the germplasm at the time of fertilization.

Similarly, in the chalcid fly, *Ageniapsis,* a chain of embryos is formed from a single egg, and these, according to Marschal, are all of the same sex.

Newman and Patterson also have shown that in the armadillo, *Tatusia,* there are customarily produced four young within a single chorion, all of which are of the same sex.

These facts point toward the conclusion that the determination of sex takes place at the time of fertilization.

5. SELECTIVE FERTILIZATION

Within the last ten years considerable evidence has been collected in support of the supposition that sex is a Mendelian character. Mendel himself, without elaborating this idea into a definite hypothesis, suggested the probability that sex is a heritable character behaving in the same way as other heritable characters.

In 1903 Castle published a paper [1] in which a tentative explanation, since abandoned, of the phenomenon of sex determination was advanced, based upon three assumptions: first, that *all germ-cells are heterozygous for sex* and, therefore, upon maturation there are formed both male and female eggs as well as male and female sperms; second, that in fertilization *the gametes always unite with their opposites so far as sex is concerned and never with their like,* with the result that each fertilized egg must carry

[1] Castle, W. E., "The Heredity of Sex." Bull. Mus. Comp. Zool., Harvard, Vol. XL, No. 4, 1903.

determiners for both sexes and be heterozygous, as indicated in Figure 58; and third, that the character of sex follows the law of *alternative dominance*, according to which in the male offspring the male determiner dominates $M(F)$, while in the female the female dominates $(M)F$.

This hypothesis is simply an attempt to explain the numerical equality of the sexes, and also the fact that the determiner for the opposite sex may be car-

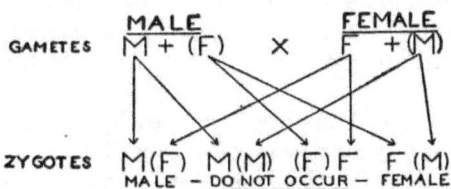

FIG. 58.—Diagram to show Castle's 1903 theory of the heredity of sex.

ried by either parent, but it leaves unanswered the question of what causes "selective fertilization" and "alternative dominance."

There appears to be some evidence that selective fertilization, which was assumed in Castle's 1903 theory, may actually occur under certain circumstances. For example, homozygous or pure yellow mice, that is, mice with a duplex determiner for yellow color, are not known. In breeding, all kinds of yellow mice behave as if heterozygous or simplex with respect to yellow color, for when any two yellow mice are bred together, they produce a certain percentage of recessives which would not happen if they were pure yellow.   In a Mendelian monohybrid cross, as has been previously pointed out, the expectation is that in the second generation one fourth of the offspring will be recessives $(DR \times DR = DD + 2\,DR + RR)$, but when yellow mice are bred together, the percentage of recessives approximates one third instead of one fourth. This apparent exception to the Mendelian ratio finds an explanation, however, when it is assumed that selective fertilization takes place in such a cross, and thus, since a D gamete never unites with another D gamete, but always with its opposite, R, pure yellow mice are unknown.

This supposition is further supported by the fact that the litters of young from yellow mice are, on an average, only three fourths as large as normal litters of mice, which is exactly what would be expected if one fourth of the possible gametic combinations (DD) fail to produce offspring.

Castle's tentative explanation of the determination of sex at least breaks away from the old conception that the sperm-cell produces male offspring and the egg-cell, females.   It agrees, too, with Darwin's idea that both sexes are present in each individual with one sex latent.   In certain parthenogenetic rotifers, aphids and daphnids, both sexes are plainly present in the female, since two kinds of easily distinguishable eggs are produced, one of which develops into males and the other into females without fertilization or any kind of a union with a sperm-cell.

6. THE NEO-MENDELIAN THEORY OF SEX

Correns (1906) avoids the difficulties of alternative dominance, which Castle's hypothesis offers, by supposing that one parent only is heterozygous with respect to sex, and this supposition is becoming more and more probable as evidence accumulates. According to this idea, there are two types

FIG. 59. — Diagram to show the neo-Mendelian theory of the heredity of sex, using sex symbols.

of cases, one when the female is the heterozygous parent and the other when the male is the heterozygous parent, as represented in Figure 59.

The formulæ for these types may be expressed in the nomenclature of the presence and absence theory,

TYPE	ZYGOTES	GAMETES
I	(Female) X O	X + O
	(Male) O O	O + O
		XO OO
		(Female) (Male)
II	(Female) X X	X + X
	(Male) X O	X + O
		XX XO
		(Female) (Male)

FIG. 60. — Diagram to show the neo-Mendelian theory of the heredity of sex according to the presence and absence hypothesis.

as follows (Fig. 60), in which the symbol x represents the female determiner in the heterozygous case of type I, and xx the female determiner when the male is the heterozygous parent.

The formulæ may be still further modified, according to Morgan, for the satisfaction of those who object to regarding the male factor as nothing positive, but simply the absence of femaleness, by assuming that a universal factor of maleness (m) is present in all cases, as shown in Figure 61.

Thus in type I of this scheme it is only when the dominant female factor F is entirely absent that maleness becomes expressed in the somatoplasm, while in type II it is necessary to have a double dose of the factor F in order to produce a female, since a single dose results in a male.

Type	Zygotes		Gametes
I	(Female) Fmfm	Fm + fm	
	(Male) fmfm	fm + fm	
		Fmfm (Female)	fmfm (Male)
II	(Female) FmFm	Fm + Fm	
	(Male) Fmfm	Fm + fm	
		FmFm (Female)	Fmfm (Male)

Fig. 61. — Diagram to show the neo-Mendelian theory of the heredity of sex with Morgan's modification, making maleness (m) present as a positive character in every gamete.

All of these three theoretical schemes agree in assuming that *one sex is heterozygous, while the other is homozygous* and that femaleness is the result of an added factor in excess of maleness.

The evidence for these conclusions has been obtained chiefly from four sources: *first*, from a microscopical examination of the germ-cells; *second*, from castration and regeneration experiments; *third*, from the results of hybridization in "sex-limited inheritance"; and *fourth*, from the behavior of hermaphrodites in heredity.

a. Microscopical Evidence

A, 1. The "X" Chromosome

In 1891 Henking called attention to the presence of two kinds of spermatozoa in the firefly, *Pyrrhocoris*, and later McClung (1901), in studying the spermatogenesis of the grasshopper, discovered a similar phenomenon with respect to the chromosomes of its spermatozoa. Soon after, Stevens and Wilson

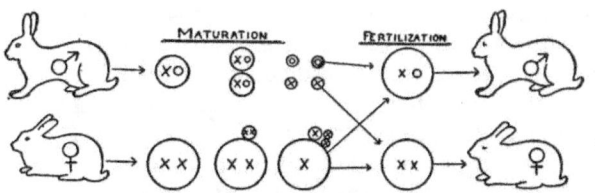

Fig. 62. — Diagram to show how numerical equality of the sexes results when one parent is homozygous (the female in this instance) and the other is heterozygous for the sex character.

working independently on various species of insects, and Boveri, on sea-urchins, found that when the male is characterized by two kinds of sperm-cells, one of which has an "extra" chromosome (the so-called "accessory" or "x" chromosome), while the other does not, the female of the same species, upon maturation of the eggs, produces mature eggs, all of which possess one "x" chromosome. The result of this heterozygous condition of the male and homozygous condition of the female with respect to the x chromosome is the theoretical equality of the sexes among the individuals formed by their union, as shown in Figure 62 or in type II of Figures 59, 60 and 61.

It will be seen that when a male gamete bearing an x chromosome unites with a female gamete also bearing an x chromosome, the outcome is a fertilized egg containing xx chromosomes. Such an egg is consequently homozygous for sex, and will develop into a female individual. In the same way when a male gamete lacking an x chromosome, as half the gametes derived from a heterozygote do, unites with a female gamete bearing an x chromosome, as all gametes from a homozygote must do, then the fertilized egg will be heterozygous, carrying only one x chromosome, and will develop into a male individual.

The chromatin difference between the two sexes may be qualitative, as Wilson holds, or quantitative, as Morgan assumes, but in either case it seems certain that, with difference in sex, there is invariably associated a definite difference in the character of the chromosomes present in the germ-cells.

These conclusions have been abundantly confirmed in various species by a large number of independent workers, and are now well established as a part of biological science. In fact, it is not at all unusual to find the technical confirmation of the x chromosome theory given as a part of the routine class work in university courses.

A, 2. Various Forms of X Chromosomes

The extra chromosome in different species may assume various forms or degrees of complexity. It may be either single or multiple. It may be paired before maturation with its absence, or with an unlike ("y") chromosome. It may be linked inseparably with some one of the ordinary chromosomes (*autosomes*), or resemble the autosomes so closely that its presence can only be assumed from analogy with other cases, and not definitely determined at all.

In all of these cases, however, there is one point of likeness, and that is that there always seems to be *additional chromatin material associated with the female sex.*

The reason for this may lie in the more highly metabolic requirements of the female, who must produce yolk or provide in some way for the maintenance of the young in addition to furnishing half of the germinal heritage.

In the microscopical evidence on this point there is one apparent exception to the rule that females are homozygous and males heterozygous with respect to sex. Baltzer (1910) found that in one of the sea-urchins an extra sex chromosome is associated with the female sex, so that two kinds of mature eggs are produced upon maturation and only one kind of sperm-cells. In other words, in this case the female is heterozygous for sex and the male homozygous, instead of the reverse which is true for all other forms thus far microscopically investigated.

Such cases as this of the sea-urchin are theoretically provided for in the formulæ under type I given above in Figures 59, 60 and 61.

P

A, 3. Sex Chromosomes in Parthenogenesis

The behavior of the chromosomes in cases of parthenogenesis, where the union of an egg-cell and a sperm-cell are not necessary for the production of a new individual, throws additional light upon the relation between chromosomes and sex determination.

For instance, among the social Hymenoptera, bees, ants, wasps, etc., the "queen" produces eggs which upon maturation, if unfertilized, develop into males or drones, all of whose cells contain a reduced amount of chromatin (Fig. 63). It is only when sexual reproduction occurs through the union of a mature egg-cell and a mature sperm-cell or spermatozoan, that the full complement of chromatin is restored to the fertilized egg and females are again produced.

Castle says: "In all known cases of parthenogenesis the female is in the duplex $(2\,n)$ condition, and the male is in the simplex (n), or partially duplex $(2\,n - 1$ condition. The female in all cases has the greater chromatin content."

b. Castration and Regeneration Experiments

Certain characters which are known as "secondary sexual characters," such as the ornamental plumage in male birds, the beard in man or the sting in worker bees, are often associated with a definite sex. When an individual is castrated, it is quite common not only for these peculiar secondary sexual characters to disappear, but also for the secondary sexual characters of

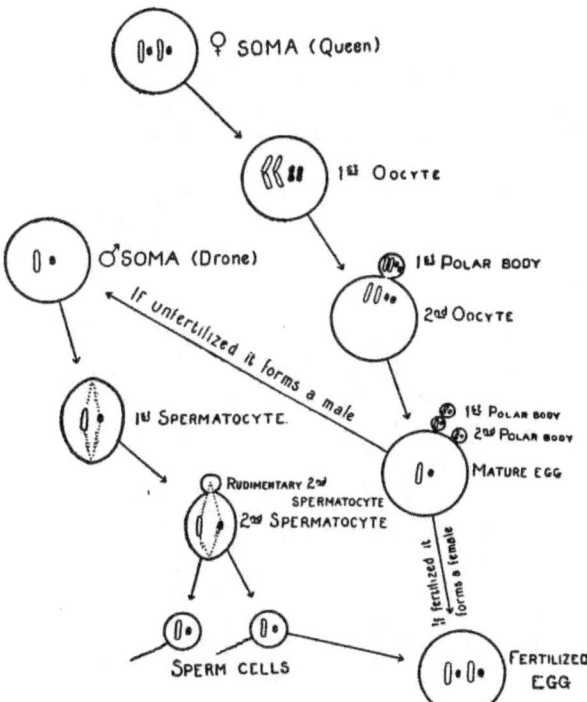

FIG. 63. — Diagram of the heredity of sex in bees, ants and wasps. The outline chromosomes represent sample somatic chromosomes. The solid black chromosomes stand for sex. The female has two sex chromosomes while the male has but one.

the opposite sex to develop to a certain degree in their stead. This indicates that the determiners for sex are intimately associated with those for the sec-

ondary sexual characters, and also that the determiners for the opposite sex are often present in a latent
condition, or, in other words, that the organism,
either male or female, is *heterozygous with respect to
sex.*

If a female of the annelid worm *Ophiotrocha*, for
example, is cut in half, it is effectually castrated, be-

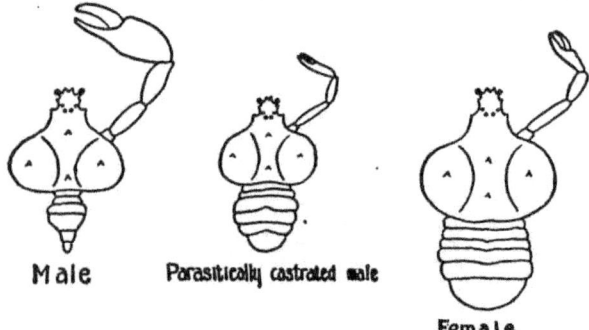

Male   Parasitically castrated male

Female

FIG. 64. — The crab, *Inachus*, parasitised by the cirripede, *Sacculina*.
Evidence that a Mendelian sex determiner is correlated with " secondary sexual characters " and that the male is heterozygous for sex
while the female is homozygous. After Smith.

cause the ovaries are in the posterior part of the body.
It has the power of regeneration, however, but when
a new posterior part is formed, it contains, not female,
but male reproductive organs. The worm is, therefore, now a male, as shown by the presence of testes
instead of ovaries, proving that it was originally
heterozygous with respect to sex. carrying one sex
latent.

According to Smith, parasitic castration is performed on the crab *Inachus*, which is found in the Bay of Naples, by a cirripede, *Sacculina*. The male crab of this species has one large claw and a narrow abdomen, while the female has no large claw, but a broad abdomen. When *Sacculina* parasitizes the female, the secondary sexual characters of the female are stunted, but not materially changed. When, on the contrary, the male is parasitized, it not only loses its distinctive large claw in subsequent molts, but it also takes on the broad abdomen of the female (Fig. 64). This apparent anomaly is quite explainable upon the assumption that the female is homozygous for sex and the accompanying secondary sexual characters, while the male is heterozygous. When maleness is destroyed in the male by the castrating parasite, therefore, the femaleness that is latent in this sex becomes manifest through the appearance of female secondary sexual characters; but when the female is castrated, no other secondary sexual characters than those already present make their appearance, since only femaleness is present in the homozygous female sex.

c. Sex-limited Inheritance

Additional evidence that sex is a character depending upon determiners which behave in Mendelian fashion is furnished by what is called sex-limited inheritance. There are certain characters known as sex-limited characters that are in no sense to be confused with secondary sexual characters which appear to be always linked with the determiner for either one sex or the other. They are, therefore, well described by the term "sex-limited."

(1) *Color-blindness*

This phenomenon may be illustrated by the inheritance of human color-blindness, a character which appears to be linked with the determiner for sex. It requires a duplex, or homozygous, dose of the determiner for color-blindness to produce a color-blind female, while only a simplex, or heterozygous, dose is

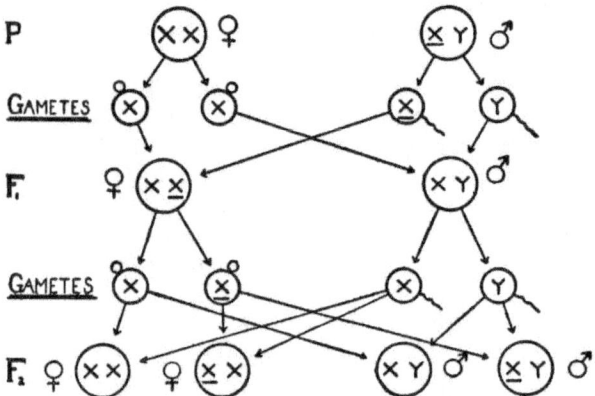

FIG. 65. — General diagram for sex-limited inheritance. The underscored symbol (X̲) represents a sex determiner with some other character (as color-blindness) linked with it.

needed to produce a color-blind male. These facts agree perfectly with the idea that the female is homozygous and the male heterozygous with respect to

sex, and that the factor for color-blindness is linked
with the determiner for sex. Sex-limited inheritance,
as shown in this case, may be illustrated by the diagram on the opposite page (Fig. 65) in which, for the
sake of simplicity, only sex chromosomes and the determiners for color-blindness are represented. Underscored ✕ represents a color-blind determiner linked
to a sex chromosome.

From this diagram, which agrees substantially
with the facts, it is apparent that a color-blind male
mated to a normal female will produce no color-blind
offspring, although the females will be "carriers"
of color-blindness, that is, will possess the factor in
simplex form and will, therefore, carry it *for the female*
in a latent condition.

The sons of such a mating having a normal mother
and a color-blind father will be absolutely free from
the defect and cannot produce color-blindness in any
of their offspring when mated with a normal strain.
If, however, the "carrier" daughters from such a
parentage, who are genotypically heterozygous for
color-blindness but phenotypically normal, mate with
normal individuals, the expectation is that one half
of the sons, and none of the daughters will be color-
blind, but that one half of these daughters will carry
the color-blind determiner in simplex form, that is, in
a condition ineffective for producing color-blindness
in female individuals.

All of the various possibilities in the inheritance of
color-blindness according to the sex-limited interpretation are indicated in the following table : —

PARENTS		EXPECTED OFFSPRING	
♂	♀	♂	♀
Normal	Color-blind	Color-blind	Carrier
Normal	Carrier	½ color-blind ½ normal	½ carrier ½ normal
Color-blind	Normal	Normal	Carrier
Color-blind	Color-blind	Color-blind	Color-blind
Color-blind	Carrier	½ color-blind ½ normal	½ color-blind ½ carrier

(2) *The English Currant-worm*

A famous case of sex-limited inheritance is that of the English currant-worm, *Abraxas*, which occurs in two varieties, viz., *Abraxas grossulariata* and *Abraxas*

Fig. 66. — *Abraxas grossulariata*, the English currant-moth, and (on the right) its paler *lacticolor* variety. From Punnett's "Mendelism."

lacticolor (Fig. 66). The lighter-colored *lacticolor* is recessive to the darker-colored *grossulariata* variety and has been found in nature associated only with the female sex.

Doncaster and Raynor, in 1908, published the results of various crosses between these two varieties which demonstrate clearly that sex is a Mendelian character and that, in this instance, maleness is homozygous and femaleness heterozygous with the determiner

	PHENOTYPE	Constitution with respect to the GROSSULARIATA factor	GENOTYPE	GAMETES
1	GROSS.♂	Duplex		
2	GROSS.♂	Simplex		
3	LACT.♂	Nulliplex		
4	GROSS.♀	Simplex		
5	LACT.♀	Nulliplex		

FIG. 67.—Key to the symbols employed in Figures 68–71. The outline symbols represent samples of the autosomes or somatic chromosomes. The black symbols stand for the "extra" or sex chromosomes. *G* above a black symbol indicates the *grossulariata* factor linked with a sex chromosome. The variety *lacticolor* occurs whenever the *grossulariata* factor is absent.

for maleness linked with the factor producing the variety *grossulariata*. A study of Figures 67–71 will make this case clear. Outline symbols represent ordinary chromosomes or autosomes, several of which are omitted for sake of clearness. The black symbols represent sex chromosomes. The letter *G* placed above a black symbol represents the *grossulariata* factor linked with a sex chromosome. The variety *lacticolor* occurs whenever the factor for *grossulariata* is absent. In this case two sex determiners are necessary to produce a male, and only one to produce a female. In the following theoretical diagrams the actual number of offspring obtained by Doncaster and Raynor in each cross is indicated outside the circles that represent the zygotes, and the parenthetical numbers refer to the five kinds of individuals catalogued in Figure 67.

In the first cross (Fig. 68) where a *lacticolor* female (5) and a *grossulariata* male (1) were bred together, the entire progeny was *grossulariata* in character with an approximate equality between the sexes, that is, 45 males (2) to 50 females (4).

When these hybrid *grossulariata* individuals, (2) and (4), were mated with each other in Cross 2 (Fig. 69), the character of *grossulariata* appeared again in both sexes, (1), (2), and (4), while the character *lacticolor* was confined as usual to females alone (5). It was only when *grossulariata* hybrid males (2) were crossed back to *lacticolor* recessive females (5) in Cross 3 (Fig. 70) that individuals of both varieties and both sexes appeared, (2), (3), (5), (4), in practically the expected equal numbers, namely, 63, 65, 70, 62. The *lacticolor* male (3) obtained by bringing together the two sex determiners necessary for maleness, each of which had been dissociated through the foregoing crosses from the sex-limited *grossulariata* factor, was entirely new to science, never having been found in nature.

CROSS 1

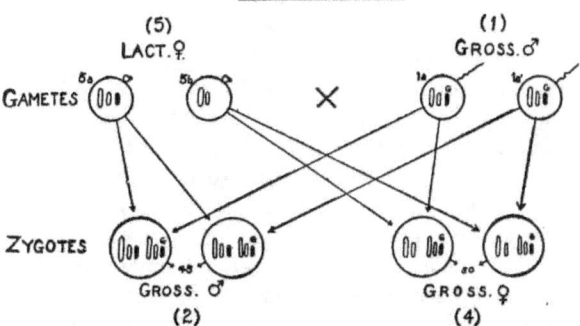

FIG. 68.—The formation of heterozygous *grossulariata* individuals, both male and female, by crossing pure *grossulariata* males with *lacticolor* females.

CROSS 2

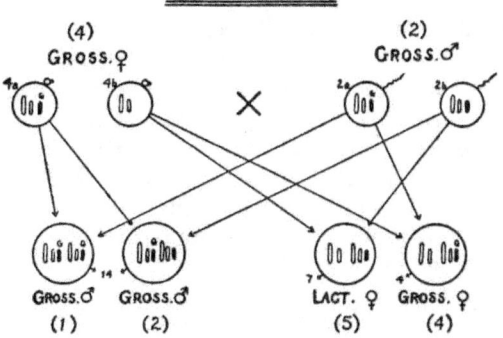

FIG. 69.—The cross-breeding of heterozygous *grossulariata* individuals.

Finally, when these newly made *lacticolor* males (3) were crossed with heterozygous *grossulariata* females (4) (Fig. 71), the proportion of sexes was approximately equal, as expected, that is, 145 males to 130 females, but all of the males were of the heterozygous *grossulariata* type (2) and all of the females of the recessive *lacticolor* type (5), showing a return to the sex-limited condition. All of these curious results find a satisfactory and complete explanation in the assumption, first, that sex is a Mendelian character carrying two determiners for maleness and one for femaleness; and, second, that the determiner for the character of *grossulariata* when present is always linked to the sex determiner.

This case is of particular interest, since it agrees with the microscopical evidence already referred to in connection with the chromosomes of Baltzer's sea-urchins, in which the male was likewise homozygous and the female heterozygous with respect to sex.

The chromosomes of *Abraxas* present certain technical difficulties which at present have not been overcome, so that we do not yet know whether the evidence of the heterozygous character of one sex and the homozygous character of the other, obtained from the breeding experiments of Doncaster and Raynor, will be confirmed upon a microscopic examination of the chromosomes in the germ-cells.

(3) *The Behavior of Hermaphrodites in Heredity*

Certain plants occur in *monœcious* form, that is, as hermaphrodites, and also in *diœcious* form, that is, with

CROSS 3

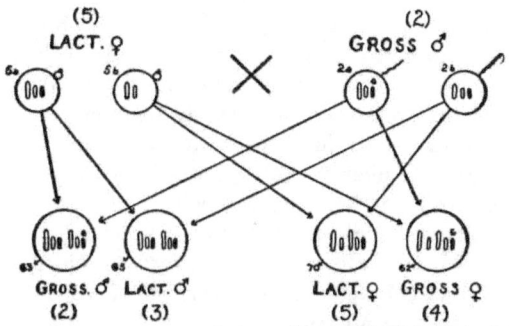

FIG. 70. — Heterozygous *grossulariata* male crossed with *lacticolor* female. One fourth of the progeny are *lacticolor* MALE, not known to occur in nature.

CROSS 4

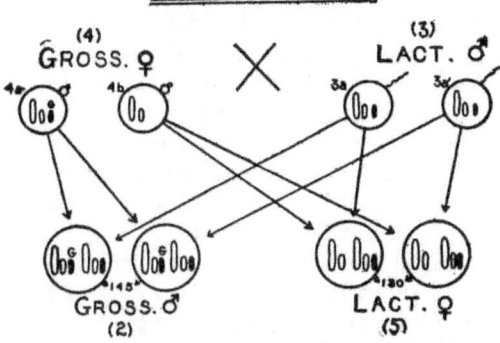

FIG. 71. — Back cross of *lacticolor* male with *grossulariata* female producing the original sex-limited condition in which all the females are of the *lacticolor* type. Data for Figures 68–71 from Doncaster and Raynor.

the sexes on separate plants. Among such dimorphic plants, *Bryonia* in particular has been investigated by Correns and *Lychnis* by Shull. Without describing the crosses made in their experiments in detail, it may be stated that when diœcious types are reciprocally crossed with hermaphroditic forms, the resulting progeny indicate plainly that one sex is homozygous while the other is heterozygous with respect to the sex character. This confirmatory evidence is quite in line with that already brought forward that sex is a Mendelian character the determiners of which are carried in the germplasm.

7. Conclusion

The evidence thus far obtainable from all sources points to the conclusion that sex is unalterably fixed at the time the egg is fertilized, by definite determiners which act in the same way as other Mendelian determiners. Dr. Shull, whose exhaustive studies in sex determination place him in the front rank as an authority on the subject, makes this conservative statement: "Nearly all the recent investigations indicate that sex is at least predominantly dependent upon the genotypic nature of the individual."

If this is so, while it furnishes the best of confirmatory evidence in support of Mendel's law, it shows that it is not possible for man to predetermine the sex of his offspring, which he has long hoped to be able to do. The following quotation from Castle may suitably close this chapter: "Negative as are the results of our study of sex control, they are perhaps not wholly without practical value. It is something to know our limitations. We may thus save time from useless attempts at controlling what is uncontrollable and devote it to more profitable employments."

CHAPTER XI

THE APPLICATION TO MAN

1. The Application of Genetics to Man

Human civilization goes hand in hand with the degree of successful interference which man exerts upon the natural forces surrounding him.

Primitive man was overwhelmed and outmastered by his environment, but civilized man harnesses nature to do his will. Savages are not proficient in the arts of cultivating plants and domesticating animals, while these are the very things upon which human progress fundamentally depends. The degree of civilization of any people is closely correlated with the degree of their success in exercising a conquering control over plants and animals. Any knowledge of the laws of heredity, therefore, as applied by man, either directly to himself or indirectly to animals and plants, is a distinct contribution to human progress.

In 1900 the National Association of British and Irish Millers, as Kellicott points out, being dissatisfied with the quality and quantity of the annual wheat yield, engaged Professor Biffen to apply his knowledge of heredity to the practical problem of improving their wheat crop. The characters desired were a short full head, beardlessness, high gluten

224

content, immunity to rust, strong supporting straw, and a high yield per acre. In the short time that has elapsed, Professor Biffen has succeeded in producing strains of wheat that combine all these desirable characters to a remarkable degree.

Such an immediate result would not have been possible before 1900, when the rediscovery of Mendel's law revolutionized man's knowledge of the action of heredity in nature.

This same knowledge which has made possible the improvement of wheat may be applied to the breeding of man, for there is no reasonable doubt that man belongs in the same evolutionary series with all other animals, as Darwin showed, and is consequently subject to the same natural laws to a considerable degree.

It must be admitted that thus far in the progress of civilization more attention has been directed to the scientific breeding of animals and plants, little as that has been, than to the scientific breeding of man. Let us hope that the future will have a different story to tell !

2. Modifying Factors in the Case of Man

There are certain qualifying factors which make the problems of genetics somewhat different in the case of man than of other organisms.

For example, mankind has come to be partially exempt from some of the natural laws that affect other organisms. Thus with respect to the workings of *natural selection* man is partially under "grace"

Q

rather than "law." Nature no longer "selects" good eyes in man by long, patient, and devious processes when poor eyes are made good almost instantly by a visit to the oculist. She has long since given up providing natural weapons of defense for those who have the wits to supply themselves more efficiently with artificial means of self-preservation, and she no longer attempts to improve the natural powers of locomotion of those who are able to tame a horse to ride upon, or who build steamships, railroads, automobiles and aeroplanes, thus accomplishing at once what would require ages at least to evolve.

Neither does the law of the *survival of the fittest* in its original sense apply equally to man and to other organisms. Human society to-day protects its unfit in hospitals, asylums, and through various philanthropies, while physicians devote themselves to the art of prolonging life beyond the period of usefulness.

We do not desire these results of our modern civilization to be otherwise, but the fact remains that some of the most inflexible and universal "natural laws" are ineffective in the case of man, and it is profitable to bear this in mind when applying the laws of genetics to man.

The laboratory for human heredity is the wide world, but it is obvious that the experimental method which has proven so effective in studying the heredity of animals and plants is impracticable in the case of man. The consideration of human heredity, therefore, must always be largely from the statistical side, consisting in an analysis of experiments already performed rather than in initiating new experiments.

Such institutions as insane asylums, prisons, sanitariums, and homes for the unfortunate are excellent foci for studying certain phases of human heredity, because they are simply convenient places where the results of similar experiments in genetics have been brought together.

3. Experiments in Human Heredity

a. The Jukes

A classic example of an experiment in human heredity which has been partially analyzed by the statistical method is that furnished by Dugdale in 1877 in the case of "Max Jukes" and his descendants. It includes over one thousand individuals, the origin of all of whom has been traced back to a shiftless, illiterate, and intemperate backwoodsman who started his experiment in heredity in western New York when it was yet an unsettled wilderness.

In 1877 the histories of 540 of this man's progeny were known, and that of most of the others was partly known. About one third of this degenerate strain died in infancy, 310 individuals were paupers who all together spent a total of 2300 years in almshouses, while 440 were physical wrecks. In addition to this, over one half of the female descendants were prostitutes, and 130 individuals were convicted criminals, including 7 murderers. Not one of the entire family had a common school education, although the children of other families in the same region found a way to educational advantages. Only 20 individuals learned a trade and 10 of these did so in state's prison.

It is estimated that up to 1877 this experiment in human breeding had cost the state of New York over a million and a quarter dollars, and the end is by no means yet in sight.

b. The descendants of Jonathan Edwards

In striking contrast to the case of Max Jukes is that of Jonathan Edwards, the eminent divine, whose famous progeny Winship describes as follows: "1394 of his descendants were identified in 1900, of whom 295 were college graduates; 13 presidents of our greatest colleges, besides many principals of other important educational institutions; 60 physicians, many of whom were eminent; 100 and more clergymen, missionaries, or theological professors; 75 were officers in the army and navy; 60 were prominent authors and writers, by whom 135 books of merit were written and published and 18 important periodicals edited; 33 American States and several foreign countries and 92 American cities and many foreign cities have profited by the beneficent influence of their eminent activity; 100 and more were lawyers, of whom one was our most eminent professor of law; 30 were judges; 80 held public office, of whom one was vice-president of the United States; 3 were United States senators; several were governors, Members of Congress, framers of state constitutions, mayors of cities, and ministers to foreign courts; one was president of the Pacific Mail Steamship Company; 15 railroads, many banks, insurance companies, and large industrial enterprises have been indebted to their management. Almost if not every department of social progress and of public weal has felt the impulse of this healthy, long-lived family. It is not known that any one of them was ever convicted of crime."

c. The Kallikak Family

A more convincing experiment in human heredity than the foregoing, since it concerns the descendants of two mothers and the same father, is furnished by the recently published history of the " Kallikak " family.[1]

During Revolutionary days, the first Martin Kallikak, — the name is fictitious, — who was descended from a long line of good English ancestry, took advantage of a feeble-minded girl. The result of their indulgence was a feeble-minded son who became the progenitor of 480 known descendants of whom 143 were distinctly feeble-minded, while most of the others fell below mediocrity without a single instance of exceptional ability.

"After the Revolutionary war, Martin married a Quaker girl of good ancestry and settled down to live a respectable life after the traditions of his forefathers. From this legal union with a normal woman there have been 496 descendants. All of

[1] "The Kallikak Family." H. H. Goddard. The Macmillan Co.

these except two have been of normal mentality and these two were not feeble-minded. . . . The fact that the descendants of both the normal and the feeble-minded mother have been traced and studied in every conceivable environment, and that the respective strains have always been true to type, tends to confirm the belief that heredity has been the determining factor in the formation of their respective characters."

4. Moral and Mental Characters behave like Physical Ones

These instances of human breeding show unmistakably that "blood counts" in human inheritance, even though the hereditary unit characters that lead to these general results have not yet been analyzed with the clearness that is possible in dealing with the characters of some animals and plants.

There is of course no question of moral and mental traits in plants, and the rôle that these play in animals is not easy to determine; but in man the case is undoubtedly much more important and complex, since mental and moral characteristics have a large share in making man what he is. There is, however, no fundamental scientific distinction which can be drawn between moral, mental, and physical traits, and they are undoubtedly all equally subject to the laws of heredity.

For instance, as an illustration of the heritability

of non-physical traits, in the Jukes pedigree three
of the daughters of Max impressed their peculiar
moral and mental characteristics in a distinctive
way upon their offspring. To quote Davenport:
"Thus in the same environment, the descendants of
the illegitimate son of Ada are prevailingly *criminal;*
the progeny of Belle are *sexually immoral;* and the
offspring of Effie are *paupers.* The difference in the
germplasm determines the difference in the prevailing
trait."

5. The Character of Human Traits

Of the mental, moral, and physical traits which
are heritable in man, some must be regarded as
generally desirable, some as indifferent, and others
as defects to be avoided if possible. In general the
majority of human traits, those which together make
up man as distinguished from other animals, do not
particularly claim the attention because they are so
universal. Some which stand out from the mass,
such as the physical traits of eye-color and the color
and character of hair, may be regarded as indifferent
so far as the welfare of the individual is concerned,
while others like skin color and certain racial features
that characterize particular strains of "blood" may,
under certain circumstances, work a social handicap
upon their possessors according to the traditions of
the community in which they appear.

A long list of desirable mental traits might be
enumerated that seem in a general way to be subject
to the laws of inheritance, although they have not

yet undergone the careful analysis demanded by modern genetics which deals in unit characters rather than in lump inheritance.

Musical, literary, or artistic ability, for example, mathematical aptitude and inventive genius, as well as a cheerful disposition or a strong moral sense are probably all gifts that come in the germplasm.

They may each be developed by exercise or repressed by want of opportunity, nevertheless they are fundamentally germinal gifts.

A genius must be born of potential germplasm. No amount of faithful plodding application can compensate for a lack of the divine hereditary spark at the start.

6. Hereditary Defects

Undesirable hereditary traits are usually defects due to the absence of some character. For instance, albinism, which occurs in several kinds of animals and also in man in one out of every 20,000 individuals (according to Elderton), is due to the absence of pigment in the skin, hair and eyes. Albinic individuals have poor eyesight because they are unable to stand strong light, being without protective pigment in the eyes. This peculiarity of albinism behaves as a recessive character both in man and in other animals. An albinic individual may, therefore, marry a normal individual without fear of producing albino children, although the children of such a mating would carry heterozygous germplasm with respect to albinism, and in cousin marriages might subsequently produce some albino children.

Davenport, in his recent work on "Heredity in Relation to Eugenics," brings together a long catalogue of human hereditary defects, although in most instances they are extremely difficult of accurate analysis. This is the case, first, because these defects so often probably depend upon a combination of determiners rather than upon a single one, and, second, because the available data are usually scattered and incomplete.

Deafness, for example, is a defect which is hereditary though exactly to what degree, it is at present impossible to state. The following table taken from the extensive work of Fay (1898) upon "Marriage of the Deaf in America" gives some idea of the results of different matings lumped together statistically.

CONDITION OF PARENTS	PERCENTAGE OF DEAF OFFSPRING
Both born deaf	25.9
One born deaf, one with acquired deafness .	6.3
One born deaf, one normal	11.9
Both with acquired deafness	2.3
One with acquired deafness, one normal . .	2.2

That two parents born deaf do not produce more than 26 per cent of deaf children is probably due to the fact, first, that each parent is in all likelihood heterozygous for deafness and that, second, the same combination of factors which is the cause of the parental defect on either side of the pedigree does not happen to recombine after segregation to form the new individual. Deafness will be produced in the offspring only when matings occur in which the proper factors are combined. Such an undesirable result is much more likely to happen if both parents come from the same, or related, hereditary strains than if they are derived from families in no way connected by blood.

Herein lies the biological objection to cousin marriage which tends to bring together, and thus to perpetuate, like defects. Outcrossing, on the contrary, through the law of dominance, tends to conceal defects and to prevent their expression.

Many other cases of human defects, such as imbecility or insanity, are extremely difficult of analysis from the standpoint of heredity because, in the first place, the defective conditions descriptively included under these vague terms are made up of a multitude of diverse conditions each of which must have a different array of determiners and, in the second place, because any one definite sort of insanity or imbecility may be conditioned by a variety of factors.

However, the difficulty of the problem is no reason for abandoning the attempt to reach its solution and to learn, if possible, "whence come our 300,000 insane and feeble-minded, our 160,000 blind or deaf, the 2,000,000 that are annually cared for by our hospitals and Homes, our 80,000 prisoners and the thousands of criminals that are not in prison,

and our 100,000 paupers in almshouses and out "
(Davenport).

7. The Control of Defects

The method of possible control of human defects
depends upon whether they are positive or negative,
that is, dominant or recessive. In those cases where
a given defect is due to a single determiner the
Mendelian expectation for the possible offspring
arising from various matings is indicated in the following table in which D stands for the defect and d
for its absence : —

The Mendelian Expectation for Defects

		If the Defect is Positive (dominant)	If the Defect is Negative (recessive)
When both parents show the defect	1	$DD \times DD$ = all DD	$dd \times dd$ = all dd
	2	$DD \times Dd = \frac{1}{2}DD + \frac{1}{2}Dd$	
	3	$Dd \times Dd = \frac{1}{4}DD + \frac{1}{2}Dd + \frac{1}{4}dd$	
When one parent only shows the defect	4	$DD \times dd$ = all Dd	$dd \times DD$ = all Dd
	5	$Dd \times dd = \frac{1}{2}Dd + \frac{1}{2}dd$	$dd \times Dd = \frac{1}{2}Dd + \frac{1}{2}dd$
When neither parent shows the defect	6		$DD \times DD$ = all DD
	7	$dd \times dd$ = all dd	$Dd \times DD = \frac{1}{2}DD + \frac{1}{2}Dd$
	8		$Dd \times Dd = \frac{1}{4}DD + \frac{1}{2}Dd + \frac{1}{4}dd$

If the defect is positive and in a duplex or homozygous condition in one parent, as in 1, 2, and 4
above, all the offspring will possess it regardless of
the germinal constitution of the other parent. In

two cases only, namely, in 3 and 5, where the defective parent is heterozygous, is there any chance of unaffected offspring, and even in these cases the defect is quite as likely to appear as not. It is obvious that the only way to rid germplasm of a dominant defect is by continued mating with recessive individuals. By this method it is possible in time to shake off the defect. When it once disappears in any individual, *it will never return* unless crossed back to a similar defective dominant strain.

In other words, such a recessive extracted from a heterozygous ancestry will breed just as true as a recessive which was pure from the start. In both instances there is an entire absence of the character in question, and it is clear that this character can thereafter never again reappear, since something cannot be derived from nothing.

On the other hand, if a defect is negative, depending upon the absence of a normal dominant determiner, as is usually the case with defects, it behaves as a Mendelian recessive, that is, it is always apparent in individuals developing from the homozygously defective germplasm.

It is certain, for example, that an imbecile which has arisen from homozygous defective germplasm carries only the determiner for imbecility in his own germplasm, and when two such recessives mate, nothing but imbecile offspring can result, for recessives breed true. Nothing plus nothing equals nothing.

An illustration of this principle is given in the following pedigree (Fig. 72) furnished by Goddard, 1910.

The result is quite different, however, when one parent only shows the defect. If the other parent is a normal homozygote, as in case 4 of the accompanying table, all the offspring will be normal in appearance, but with the bar sinister of defectiveness in their germplasm, while if the other parent is heterozygous (Case 5), one half of the progeny will be defective.

Finally, when neither parent shows defectiveness

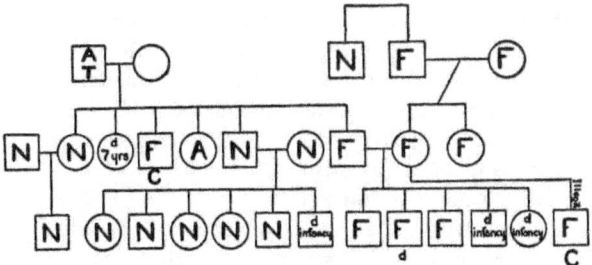

Fig. 72. — Pedigree chart illustrating the law that two defective parents have only defective offspring. *A*, alcoholic; *C*, criminalistic; *d*, died; *F*, feeble-minded; *T*, tubercular. After Goddard.

but one carries the defect as a heterozygote (Case 7), then there will be no defective children, while if both parents are heterozygous there is one chance in four that the offspring will be defective.

As a matter of fact, defectives usually mate with defectives for the simple reason that normals ordinarily avoid them, so it comes about that streams of poor germplasm naturally flowing together tend to the inbreeding of like defects.

Davenport[1] lays down the following general eugenic rules for the guidance of those who would produce offspring wisely: "If the negative character is, as in polydactylism and night-blindness, the *normal* character, the normals should marry normals, and they may be even cousins. If the negative character is *abnormal*, as imbecility and liability to respiratory diseases, then the marriage of two abnormals means probably all children abnormal; the marriage of two normals from defective strains means about one quarter of the children abnormal; but the marriage of a normal of the defective strain with one of a normal strain will probably lead to strong children. The worst possible marriage in this class of cases is that of cousins from the defective strain, especially if one or both have the defect. In a word, the consanguineous marriage of persons one or both of whom have the same undesirable defect, is highly unfit, and the marriage of even unrelated persons who both belong to strains containing the same undesirable defect is unfit. Weakness in any characteristic must be mated with strength in that characteristic; and strength may be mated with weakness."

8. INBREEDING

The whole matter of inbreeding and the part it plays in emphasizing defects has received a fresh interpretation in the light of Mendelism.

There is a widespread popular belief that inbreeding is injurious and that it is necessary to outcross

---

[1] Davenport. *Rep. of Amer. Breeders' Assoc.*, Vol. VI, p. 431, 1910.

in order to maintain the vigor and avoid the defects of any line. In the case of mankind, consanguineous marriage of various degrees has long been forbidden by law or custom in many races, particularly among the Jews, Mohammedans, Indians and Romans. On the other hand, the Persians, Greeks, Phœnicians and Arabs have freely practised inbreeding, while one of the longest of known human pedigrees, a royal line of Egypt, was notorious for close inbreeding, even to the mating of brother and sister.

There has been a greater degree of inbreeding in the Puritan stock of New England than is commonly realized. David Starr Jordan points out that a child of to-day, supposing no inbreeding of relatives had occurred, would have had in the time of William the Conqueror, thirty generations ago, 8,598,094,592 living ancestors. If this theoretical supposition were really so, it would seem quite possible for every New Englander to-day to have had at least one ancestral representative who won glory under William.

The difference between the unthinkable number given above and the actual number of probable ancestors alive thirty generations ago emphasizes the fact that inbreeding must have occurred freely.

There are, indeed, various well-known provisions in nature to insure inbreeding. The majority of plants are probably self-fertilized while hermaphroditic animals, which sometimes at least are self-fertilized particularly among the lower forms, are very common.

Nature has secured, on the other hand, often by

elaborate devices, a separation of the sexes, especially among the higher organisms, and in consequence there has arisen an unavoidable necessity of out-crossing. The intricate adaptations existing between insects and flowers, for example, seem to be directed entirely toward insuring outcrossing among plants.

9. Experiments to test the Effect of In-breeding

Numerous experiments to test the effect of inbreeding have been carried out upon various organisms.

Darwin, for instance, planted morning-glories, *Ipomœa*, derived from the same stock of seeds, in two beds which were laid out side by side, that is, in an environment as nearly the same as possible, but with half of the beds screened from insects which usually transfer pollen from flower to flower. In the screened half where all insects were excluded the flowers were of necessity self-fertilized, while in the exposed half they were presumably cross-pollinated by the insects which had free access to them. The seeds produced in the two beds were kept separate and the experiment was continued for ten years, so that at the end of that time two lots of morning-glories, one self-fertilized for ten generations and the other presumably cross-pollinated for the same length of time, were obtained for comparison. The criterion Darwin used was the vigor of the plants as shown by the length of the vine. He found that the cross-pollinated plants were to the self-pollinated ones as 100 to 53, and his conclusion was consequently, that cross-pollination is beneficial and self-pollination is detrimental.

Ritzema-Bos inbred rats for twenty generations. For the first ten generations the average number of young per litter was 7.5, while for the last ten generations it fell to 3.2.

Weismann inbred mice for twenty-nine generations and obtained a parallel result. For the first ten generations the average number per litter was 6.1, for the second ten generations 5.6, and for the last nine generations 4.2.

Shull found in growing Indian corn that loss of vigor results from continual self-fertilization, and many breeders have had similar experiences with other plants and animals.

On the other hand, in the case of the pomace fly, *Drosophila*, Castle inbred brother and sister for fifty-nine generations without diminishing the fertility of the line. No arbitrary law with respect to the effects of inbreeding upon vigor and fertility can be laid down, therefore, which will apply equally to all cases.

10. THE INFLUENCE OF PROXIMITY

Inbreeding is often the result of proximity. Insular or isolated communities, slums in cities, where those of one language herd together, or hovels in the backwoods, where degenerates of a kind are kept in intimate association, as well as asylums of various

R

sorts in which similar defectives are promiscuously
housed under the same roof, are all potent agencies
to insure human inbreeding.

Similarly, localities which have been devastated
by migrations of the most effective blood, as, for
example, parts of Ireland or many rural villages
in New England, are frequently characterized by a
population showing a large percentage of defectiveness. The able-bodied and ambitious go forth into
the world to seek their fortunes, while the deficient
in body or spirit are left behind where, under the
spell of proximity, they perpetuate their deficiencies.

The part that improved transportation has played
in mixing up populations and in counteracting the
effects of stagnation on human heredity, through inbreeding under the inertia of proximity is very great.
Before the days of railroads, cousin-marriages were
much more frequent than they are now.

From a biological point of view there is something
to be said in favor of the rape of the Sabines in the
past and for the pursuit of American heiresses by
European nobility to-day.

11. Inbreeding in the Light of Mendelism

Inbreeding in itself may not necessarily be injurious. The consequence of inbreeding as shown by
the working of Mendelian laws is that latent or
recessive characters tend to become homozygous and
so brought to the surface, while outcrossing brings
about the formation of heterozygous traits which

mask recessive characters and render them ineffective.

Cousin-marriages, although producing a high percentage of defects, do not necessarily reproduce undesirable traits. They simply bring out latent or recessive characters for the reason that under these conditions defect meets defect instead of the opposite normal condition which would dominate the defect and cause it not to appear.

Since a recessive trait is properly regarded as the *absence* of a positive dominant character, it more frequently stands for an undesirable feature than otherwise. Thus it comes about that inbreeding, by combining negative features, may "produce" a defective strain.

Outcrossing always increases heterozygous combinations in the germplasm and covers up undesirable recessive traits through the introduction of additional dominant traits. Inbreeding, on the contrary, tends to simplify the germplasm, that is, to make it more homozygous, and so to bring recessive defects to the surface.

CHAPTER XII

HUMAN CONSERVATION

1. How Mankind may be Improved

THERE are two fundamental ways to bring about human betterment, namely, by improving the individual and by improving the race. The first method consists in making the best of whatever heritage has been received by placing the individual in the most favorable environment and developing his capacities to the utmost through education. The second method consists in seeking a better heritage with which to begin the life of the individual. The first method is immediate and urgent for the present generation. The second method is concerned with ideals for the future, and consequently does not usually present so strong an appeal to the individual.

The first is the method of *euthenics*, or the science of learning to live well. The second is *eugenics*, which Galton defines as "the science of being well born."

These two aspects of human betterment, however, are inseparable. Any hereditary characteristic must be regarded, not as an independent entity, but as *a reaction between the germplasm and its environment.* The biologist who disregards the fields of educational endeavor and environmental influence, is equally at

244

fault with the sociologist who fails sufficiently to realize the fundamental importance of the germplasm.

Without euthenic opportunity the best of heri-
tages would never fully come to its own. Without
the eugenic foundation the best opportunity fails
of accomplishment. The euthenic point of view,
however, must not distract the attention now, for
the present chapter is particularly concerned with
the program of eugenics.

2. MORE FACTS NEEDED

Since the point of attack in human heredity must
be largely statistical, it is of the first importance to
collect more facts. Our actual knowledge is confused with a mass of tradition and opinion, much of
which rests upon questionable foundations. The
great present need is to learn more facts; to sift the
truth from error in what is already known; and to
reduce all these data to workable scientific form.
Much progress is being made in this direction, owing
to the impetus given by the revival of Mendel's
illuminating work, but as yet the science of eugenics
is in its infancy.

The most systematic and effective attempt in this
country to collect reliable data concerning heredity
in man has been initiated by the Eugenics Section
of the American Breeders' Association under the
secretaryship of Dr. C. B. Davenport. In 1910 the
Eugenics Record Office, with a staff of expert field
and office workers and an adequate equipment of
fire-proof vaults, etc., for the preservation of records,

was opened at Cold Spring Harbor, Long Island, New
York, with Mr. H. H. Laughlin as superintendent.
"The main work of this office is investigation into
the laws of inheritance of traits in human beings and
their application to eugenics. It proffers its serv-
ices free of charge to persons seeking advice as to
the consequences of proposed marriage matings. In
a word, it is devoted to the advancement of the
science and practice of eugenics." The publication of results from the Eugenics Record Office has
already been begun.

The Volta Bureau, founded about twenty-five
years ago in Washington by Dr. Alexander Graham
Bell, is collecting data with reference to deafness
and has now systematically arranged particulars concerning the history of over 20,000 individuals. In
England, also, the Galton Laboratory for Eugenics,
founded in 1905, is systematically collecting facts
about human pedigrees and publishing the results in
a compendious "Treasury of Human Inheritance."

Besides these special bureaus of investigation,
innumerable facts about the inheritance of particular
traits are being incidentally brought together and
made available in various institutions and asylums
throughout the world which are immediately concerned with the care of defectives of different types.
It is in connection with such institutions for defectives that much of the most successful "field work"
of the Eugenics Section of the American Breeders'
Association is being accomplished in the United
States.

3. FURTHER APPLICATION OF WHAT WE KNOW NECESSARY

Human performance always lags behind human knowledge. Many persons who are fully aware of the right procedure do not put their knowledge into practice. It follows, therefore, that any program of eugenics which does not grip the imagination of the common people in such a way as to become an effective part of their very lives is bound to remain largely an academic affair for utopians to quarrel and theorize over.

It is not enough to collect facts and work out an analysis and interpretation of them, for, important as this preliminary step is, it must be followed by a convincing campaign of education.

The lives of the unborn do not force themselves upon the average man or woman with the same insistency as the lives already begun. In the midst of the overwhelming demands of the present, the appeal of posterity for better blood is vague and remote. If every individual regarded the germplasm he carries as a sacred trust, then it would be the part of an awakened eugenic conscience to restrain that germplasm when it is known to be defective or, when it is not defective, to hand it on to posterity with at least as much foresight as is exercised in breeding domestic animals and cultivated plants.

The eugenic conscience is in need of development, and it is only when this becomes thoroughly aroused in the rank and file of society as well as among the

leaders, that a permament and increasing betterment of mankind can be expected.

4. THE RESTRICTION OF UNDESIRABLE GERM-PLASM

A negative way to bring about better blood in the
world is to follow the clarion call of Davenport and
"dry up the streams that feed the torrent of defective and degenerate protoplasm." This may
be partially accomplished, at least in America, by
employing the following agencies: control of immigration; more discriminating marriage laws; a
quickened eugenic sentiment; sexual segregation of
defectives; and finally, drastic measures of asexualization or sterilization when necessary.

a. Control of Immigration

The enforcement of immigration laws tends to
debar from the United States not only many undesirable individuals, but also incidentally to keep out
much potentially bad germplasm that, if admitted,
might play havoc with future generations.

For example, during the year of 1908, 65 idiots,
121 feeble-minded, 184 insane, 3741 paupers, 2900
individuals having contagious diseases, 53 tuber-
culous individuals, 136 criminals, and 124 prostitutes
were caught in the sieve at Ellis Island alone and
turned back from this country by the immigration
officials. These 7000 and more individuals probably
were the bearers of very little germplasm that we
are nationally not better off without.

Eugenically, the weak point in the present application of immigration laws is that criteria for exclusion are phenotypic in nature rather than genotypic, and consequently much bad germplasm comes through our gates hidden from the view of inspectors because the bearers are heterozygous, wearing a cloak of desirability over undesirable traits.

It is not enough to lift the eyelid of a prospective parent of American citizens to discover whether he has some kind of an eye-disease or to count the contents of his purse to see if he can pay his own way. The official ought to know if eye-disease runs in the immigrant's family and whether he comes from a race of people which, through chronic shiftlessness or lack of initiative, have always carried light purses.

In selecting horses for a stock-farm an expert horseman might rely to a considerable extent upon his judgment of horseflesh based upon inspection alone, but the wise breeder does more than take the chances of an ordinary horse trader. He wants to be assured of the *pedigree* of his prospective stock. It is to be hoped that the time will come when we, as a nation, will rise above the hazardous methods of the horse trader in selecting from the foreign applicants who knock at our portals, and that we will exercise a more fundamental discrimination than such a haphazard method affords, by demanding a knowledge of the germplasm of these candidates for citizenship, as displayed in their pedigrees.

This may possibly be accomplished by having trained inspectors located abroad in the communi-

ties from which our immigrants come, whose duty
it shall be to look up the ancestry of prospective
applicants and to stamp desirable ones with approval.
The national expense of such a program of genealogi-
cal inspection would be far less than the maintenance of introduced defectives, in fact it would
greatly decrease the number of defectives in the
country.   At the present time this country is spending
over one hundred million dollars a year on defectives
alone, and each year sees this amount increased.

The United States Department of Agriculture
already has field agents scouring every land for
desirable animals and plants to introduce into this
country, as well as stringent laws to prevent the im-
portation of dangerous weeds, parasites, and organisms of various kinds.   Is the inspection and super-
vision of human blood less important?

b. More Discriminating Marriage Laws

Every people, including even the more primitive
races, make customs or laws that tend to regulate
marriage.   Of these, the laws which relate to the
eugenic aspect of marriage are the only ones that
concern us in this connection.   "Marriage," says
Davenport, "can' be looked at from many points of
view.   In novels as the climax of human courtship;
in law largely as two lines of property descent;   in
society, as fixing a certain status; but in eugenics,
which considers its biological aspect, marriage is an
experiment in breeding."

Certain of the United States have laws forbidding the marriage of epileptics, the insane, habitual drunkards, paupers, idiots, feeble-minded, and those afflicted with venereal diseases. It would be well if such laws were not only more uniform and widespread, but also more rigidly enforced.

It is quite true that marriage laws in themselves do not necessarily control human reproduction, for illegitimacy is a factor that must always be reckoned with; nevertheless such laws do have an important influence in regulating marriage and consequent reproduction.

Marriage laws may, however, sometimes bring about a deplorable result eugenically, as in the case of forced marriage of sexual offenders in order to legalize the offense and "save the woman's honor." To compel, under the guise of legality, two defective streams of germplasm to combine repeatedly and thereby result in defective offspring just because the unfortunate event happened once illegitimately, is fundamentally a mistake. Darwin says: "Except in the case of man himself hardly any one is so ignorant as to allow his worst animals to breed."

c. An Educated Sentiment

A far more effective means of restricting bad germplasm than placing elaborate marriage laws upon our statute-books is to educate public sentiment and to foster a popular eugenic conscience, in the absence of which the safeguards of the law must forever be largely without avail.

Such a sentiment already generally exists to a

large extent with respect to incest, and the marriage
of persons as noticeably defective as idiots or those
afflicted with insanity, and also in America with respect to miscegenation, but a cautious and intelligent
examination of the more obscure defective traits,
exhibited in the somatoplasms of the various members of families in question, is largely an ideal of the
future. Under existing conditions non-eugenic con-
siderations such as wealth, social position, etc., often
enter into the preliminary negotiations of a marriage
alliance, but an equally unromantic caution with
reference to the physical, moral, and mental characters that make up the biological heritage of con-
tracting parties is less usual.

The scientific attitude is not necessarily opposed
to the romantic way of looking at things. Science
is simply "organized common sense," and romance, ·
that dispenses with this balance-wheel, although it
may be entertaining and always exciting at first,
is sure to be disappointing in the end. Marriages
may be "made in heaven," but, as a matter of fact,
children are born and have to be brought up on earth.
It follows without saying that it will be much easier
to stamp out bad germplasm when an educated sentiment becomes common among all people everywhere.

d. Segregation of Defectives

Persons with hereditary defects, such as epileptics,
idiots, and certain criminals, who become wards of
the state, should be segregated so that their germplasm may not escape to furnish additional burdens

to society. "We have become so used to crime, disease and degeneracy that we take them for necessary evils. That they were in the world's ignorance, is granted. That they must remain so, is denied" (Davenport).

"The great horde of defectives once in the world have the right to live and enjoy as best they may whatever freedom is compatible with the lives and freedom of other members of society," says Kellicott, but society has a right to protect itself against repetitions of hereditary blunders.

There is one grave danger connected with the administration of our humane and commendable philanthropies toward the unfortunate, for it frequently happens that defectives are kept in institutions until they are sexually mature or are partly self-supporting, when they are liberated only to add to the burden of society by reproducing their like.

Furthermore, if defectives of the same sort are collected together in the same institutions, unless sexual segregation is strictly maintained, they may by the very circumstance of proximity tend to reproduce their kind just as defectives in any isolated community tend to multiply.

David Starr Jordan cites the interesting case of *cretinism* which occurs in the valley of Aosta in northern Italy, to prove the wisdom of the sexual segregation of defectives. Cretinism is an hereditary defect connected with an abnormal development of the thyroid gland which results in a peculiar form of idiocy usually associated with goitre.

"In the city of Aosta the goitrous *cretin* has been for centuries an object of charity. The idiot has received generous support, while the poor farmer or laborer with brains and no goitre has had the severest of struggles. In the competition of life a premium has thus been placed on imbecility and disease. The *cretin* has mated with *cretin*, the goitre with goitre, and charity and religion have presided over the union. The result is that idiocy is multiplied and intensified. The *cretin* of Aosta has been developed as a new species of man. In fair weather the roads about the city are lined with these awful paupers — human beings with less intelligence than a goose, with less decency than the pig."

Whymper, writing in 1880, further observes: "It is strange that self-interest does not lead the natives of Aosta to place their *cretins* under such restrictions as would prevent their illicit intercourse; and it is still more surprising to find the Catholic Church actually legalizing their marriage. There is something horribly grotesque in the idea of solemnizing the union of a brace of idiots, and, since it is well known that the disease is hereditary and develops in successive generations the fact that such marriages are sanctioned is scandalous and infamous."

Since 1890 the *cretins* have been sexually segregated, and in 1910 Jordan reported that they were nearly all gone.

e. Drastic Measures

A fifth method of restricting undesirable germplasm in the case of confirmed criminals, idiots, imbeciles, and rapists may be mentioned, namely, the extreme treatment of either asexualization or vasectomy. The latter is a minor operation confined to the male which occupies only a few moments and requires at most only the application of a local anæsthetic, such as cocaine. There are no disturbing or even inconvenient after effects from this operation. It consists in removing a small section of each sperm duct and is entirely effectual in preventing subsequent parenthood.

In the female the corresponding operation, which consists in removing a portion of each Fallopian tube, is much more severe, but not impracticable or dangerous.

Eleven states [1] already have sterilization laws providing for certain cases and "could such a law be enforced in the whole United States, less than four generations would eliminate nine tenths of the crime, insanity and sickness of the present generation in our land. Asylums, prisons and hospitals would decrease, and the problems of the unemployed, the indigent old, and the hopelessly degenerate would cease to trouble civilization."

5. THE CONSERVATION OF DESIRABLE GERMPLASM

Not only negatively by the restriction of undesirable germplasm, but also positively by the conservation of desirable germplasm, may the eugenic ideal be approached.

[1] Indiana, 1907; Washington, 1909; California, 1909; Connecticut, 1909; Nevada, 1911; Iowa, 1911; New Jersey, 1911; New York, 1912; No. Dakota, 1913; Michigan, 1913; Kansas, 1913.

It is possible that if some of the philanthropic endeavor now directed toward alleviating the condition of the unfit should be directed to *enlarging the opportunity of the fit*, greater good would result in the end. In breeding animals and plants the most notable advances have been made by isolating and developing the best, rather than by attempting to raise the standard of mediocrity through the elimination of the worst.

One leader is worth a score of followers in any community, and the science of genetics surely gives to educators the hint that it is wiser to cultivate the exceptional pupil who is often left to take care of himself than to expend all the energies of the instructor in forcing the indifferent or ordinary one up to a passing standard. The campaign for human betterment in the long run must do more than *avoid mistakes*. It must become aggressive and take advantage of those human mutations or combinations of traits which appear in the exceptionally endowed.

There are various ways in which this improvement of society may be brought about.

a. By Subsidizing the Fit

The following unconfirmed newspaper clipping illustrates the point of what is meant by subsidizing the fit so far as certain physical characteristics are concerned. "Berlin. Dec. 11, 1911. The Emperor is reported to be interested in a plan proposed by Professor Otto Hauser for the propagation of a fixed German type of humanity, — a type which

will be as fixed as the Jewish in its characteristics,
if the suggestions of the professor can ever be carried
out.   The fixed type is to be produced as follows : —
Only 'typical' couples are to be allowed to mate.
The man is to be not more than thirty years old,
the woman not over twenty-eight, and each have a
perfect health certificate. The man should be at
least five feet seven inches tall; the woman not under
five feet six inches.   Neither the man nor the woman
should have dark hair. Its tint may range from
blonde to auburn. The eyes of the pair should be
pure blue without any tint of brown. The complexion should be fair to ruddy without any suggestion of
heaviness or 'beefiness.' The nose ought to be
strong and narrow, the chin square and powerful, and
the skull well developed at the back. The man and
the woman must be of German descent and must
bear a German name and speak the language of
Germany. These 'mated couples' are to get a
wedding gift of $125 and an additional grant for
each child born. The couples may settle in the
United States if they prefer." This reported attempt to establish a Prussian type of "Hauser
blondes" at least points the way to one sort of a
positive eugenic method that might possibly be
employed with respect to certain physical characteristics.

It should be remembered, however, that the
eugenic ideal is not by any means confined to physical traits alone.

s

b. By Enlarging Individual Opportunity

Much good human germplasm goes to waste through ineffectiveness on account of unfavorable environment or lack of a suitable opportunity to develop.

Every agency which contributes toward increasing the opportunity of the individual to attain to a better development of his latent possibilities is in harmony with a thoroughly positive eugenic practice. Thus better schools, better homes, better living conditions, in short, all euthenic endeavor, directly serves the eugenic ideal by making the best out of whatever germinal equipment is present in man.

c. By Preventing Germinal Waste

Much good protoplasm fails to find expression in the form of offspring because one or the other of possible parents is cut off either by preventable death or by social hindrances. To avoid such calamities is a part of the positive program of eugenics.

1. Preventable Death

War, from the eugenic point of view, is the height of folly, since presumably the brave and the physically fit march away to fight, while in general the unqualified stay at home to reproduce the next generation. When a soldier dies on the battlefield or in the hospital, it is not alone a brave man who is cut off, but it is the termination of a probably desirable strain of germplasm. The Thirty Years' War in

Germany cost 6,000,000 lives, while Napoleon in his campaigns drained the best blood of France.

David Starr Jordan has presented this matter very clearly. He points out that the "man with a hoe" among the European peasantry is not the result of centuries of oppression, as he has been pictured, but rather the dull progeny resulting from generations of the unfit who were left behind when the fit went off to war never to return.

Benjamin Franklin, with characteristic wisdom, sums up the situation in the following epigram: "Wars are not paid for in war time; the bill comes later."

2. Social Hindrances

There are many conditions of modern society which act non-eugenically.

For instance, the increasing demands of professional life prolong the period necessary for preparation, which, with the "cost of high living," tends toward late marriage. In this way much of the best germplasm is very often withheld from circulation until it is too late to be effective in providing for the succeeding generation.

Certain occupations such as school-teaching and nursing by women are filled by the best blood obtainable, yet this blood is denied a direct part in molding posterity, since marriage is either forbidden or regarded as a serious handicap in such lines of work. Advertisements concerning "unincumbered help" and "childless apartments" tell their own deplorable tale.

One of the darkest features of the dark ages from a eugenic standpoint was the enforced celibacy of the priesthood, since this resulted, as a rule, in withdrawing into monasteries and nunneries much of the best blood of the times, and this uneugenic custom still obtains in many quarters to-day.

6. Who shall sit in Judgment?

In the practical application of a program of eugenics there are many difficulties, for who is qualified to sit in judgment and separate the fit from the unfit?

There are certain strongly marked characteristics in mankind which are plainly good or bad, but the principle of the independence of unit characters demonstrates that no person is wholly good or wholly bad. Shall we then throw away the whole bundle of sticks because it contains a few poor or crooked ones?

The list of weakling babies, for instance, who were apparently physically unfit and hardly worth raising upon first judgment, but who afterwards became powerful factors in the world's progress, is a notable one and includes the names of Calvin, Newton, Heine, Voltaire, Herbert Spencer and Robert Louis Stevenson.

Or, take another example. Elizabeth Tuttle, the grandmother of Jonathan Edwards whose remarkable progeny was referred to in a preceding chapter, is described as "a woman of great beauty, of tall and commanding appearance, striking carriage, of strong will, extreme intellectual vigor and mental grasp akin to rapacity," but *with an extraordinary deficiency in moral sense.* She was divorced from her husband "on the ground of adultery and other immoralities. . . . The evil trait was in the blood, for one of her sisters murdered her own son and a brother murdered his own sister." That Jonathan Edwards owed his remarkable qualities largely to his grandmother rather than to his grandfather is shown by the fact that Richard Edwards, the grandfather, married again after his divorce and had five sons and one daughter, but none of their numerous progeny "rose above mediocrity, and their descendants gained no abiding reputation." As shown by subsequent events, it would have been a great eugenic mistake to have deprived the world of Elizabeth Tuttle's germplasm, although it would have been easy to find judges to condemn her.

Dr. C. V. Chapin recently said with reference to the eugenic regulation of marriage by physician's certificate: "The causes of heredity are many and very conflicting. The subject is a difficult one, and I for one would hesitate to say, in a great many cases where I have a pretty good knowledge of the family, where marriage would, or would not, be desirable."

Desirability and undesirability must always be regarded as relative terms more or less indefinable. In attempting to define them, it makes a great difference whether the interested party holds to a puritan or a cavalier standard. To show how far human judgment may err as well as how radically human opinion changes, there were in England, as recently as 1819, 233 crimes punishable by death according to law.

One needs only to recall the days of the Spanish Inquisition or of the Salem witchcraft persecution to realize what fearful blunders human judgment is capable of, but it is unlikely that the world will ever see another great religious inquisition, or that in applying to man the newly found laws of heredity there will ever be undertaken an equally deplorable eugenic inquisition.

It is quite apparent, finally, that although great caution and broadness of vision must be exercised in bringing about the fulfilment of the highest eugenic ideals, nevertheless in this direction lies the future path of human achievement.

BIBLIOGRAPHY

A few recent works of a general nature are listed below. Several of these books contain bibliographies of technical papers and other original sources of information.

For the past five years *The American Naturalist* has been particularly devoted to papers on the problems of heredity.

– BATESON, W. 1909. Mendel's Laws of Heredity. Cambridge University Press. Cambridge.

BAUR, E. 1911. Einführung in die experimentelle Vererbungslehre. Gebrüder Borntraeger. Berlin.

– CASTLE, W. E. 1911. Heredity in Relation to Evolution and Animal Breeding. D. Appleton and Co. New York.

– COULTER, CASTLE, DAVENPORT, EAST, and TOWER. 1912. Heredity and Eugenics. University of Chicago Press. Chicago.

– DAVENPORT, C. B. 1911. Heredity in Relation to Eugenics. Henry Holt and Co. New York.

GALTON, F. 1889. Natural Inheritance. Macmillan and Co. London.

GODDARD, H. H. 1912. The Kallikak Family. The Macmillan Co. New York.

CORRENS, C. 1912. Die neuen Vererbungsgesetze. Gebrüder Borntraeger. Berlin.

DARBISHIRE, A. D. 1912. Breeding and the Mendelian Discovery. Cassell and Co., Ltd. London.

EAST, E. M. 1907. The Relation of Certain Biological Principles to Plant Breeding. Bull. 158. Conn. Agric. Exp. Sta.

GODLEWSKI, E. 1909. Das Vererbungsproblem im Lichte dei Entwicklungsmechanik. Engelmann. Leipzig.

GOLDSCHMIDT, R. 1911. Einführung in die Vererbungswissenschaft. Engelmann. Leipzig.

JOHANNSEN, W. 1909. Elemente der exakten Erblichkeitslehre. Fischer. Jena.

HAECKER, V. 1912. Allgemeine Vererbungslehre. Vieweg und Sohn. Braunschweig.

KELLICOTT, W. E. 1911. The Social Direction of Human Evolution. D. Appleton and Co. New York.

LOCK, R. H. 1909. Recent Progress in the Study of Variation, Heredity and Evolution. Murray. London.

LOTSY, J. P. 1906-1908. Vorlesungen über Descendenztheorien. Fischer. Jena.

MORGAN, T. H. 1907. Experimental Zoology. The Macmillan Co. New York.

MONTGOMERY, T. H. 1906. The Analysis of Racial Descent in Animals. Henry Holt and Co. New York.

PUNNETT, R. C. 1911. Mendelism. The Macmillan Co. New York.

REID, ARCHDALL. 1905. The Principles of Heredity. Chapman and Hall. London.

THOMSON, J. A. 1908. Heredity. Murray. London.

WATSON, J. A. S. 1912. Heredity. The Dodge Publishing Co. New York.

WEISMANN, A. 1904. The Evolution Theory. London.

SCHALLMAYER, W. 1910. Vererbung und Auslese im Lebenslauf der Völker. Fischer. Jena.

DE VRIES, H. 1905. Species and Varieties, their Origin by Mutation. The Open Court Publishing Co. Chicago.

INDEX

Abnormal fertilization, 30.
Abraxas, 216.
Acquired callosities, 91.
Acquired characters, definition of, 78.
Ageniapsis, 202.
Agouti, 163, 171.
Albinism, 59, 151, 232.
Alcoholism, 93.
Alternative inheritance, 121.
Amblystoma, 90, 129.
American Breeders' Assoc., 245.
Amitosis, 20.
Ammonites, 71.
Amphimixis, 55, 81.
Anaphase, 20.
Ancon sheep, 68.
Andalusian fowl, 175, 180.
Angora hair, 129.
Antirrhinum, 173.
Ants, 210.
Aphids, 204.
Appendix, vermiform, 150.
Arithmetical mean, 45.
Armadillo, 202.
Arrested development, 149.
Artistic ability, 232.
Ascaris, 11, 17.
Asexualization, 255.
Asexual spores, 9.
Atavism, 146.
Autosomes, 209.
Average deviation, 45.
Axolotl, 90.
Azaleas, double, 63.

BALLS, 129.
BALTZER, 209, 220.
Banana fly (see Drosophila).
Banded shell, 129.
Barley, 129, 154.
Barrier of Linnæus, 62.

BATESON, 2, 41, 55, 69, 124, 129, 130, 133, 149, 160, 161, 171, 173, 180, 182.
Battle scars, 87.
BAUR, 129, 130, 173, 180.
Beans, 102, 115.
Beardless barley, 129.
Beech leaves, 43, 49.
Beech, purple, 63.
BEECHER, 70.
Bees, 210.
Begonia, 7.
Belemnites, 71.
Belgian hare, 183, 193.
BELL, 246.
Beneden, van, 21, 23.
Bertillon system of identification, 38.
BIFFEN, 129, 224.
Bimodal polygons, 48.
Biological inheritance, 75.
Biometry, 42.
Birth-marks, 87.
Blending inheritance, 121, 174, 182.
Blonde type, 257.
BLUMENBACH, 197.
Boleyn, Anne, 59.
Booted poultry, 182.
BORN, 200.
BOVERI, 11, 17, 18, 24, 25, 32, 207.
Brachydactyly, 129.
Branched habit, 129.
BROOKS, 75.
Bryonia, 222.
Bryozoa, 8.
BURBANK, 156.

Calvin, 260.
Canaries, 129.
Capsella, 89.
Captivity, effect of, 152.
Carnations, double, 63.
Carriers of color-blindness, 215.

Printed in the United States of America.

THE following pages contain advertisements of a few of the Macmillan books on kindred subjects

The Science of Human Behavior

Biological and Psychological Foundations

By MAURICE PARMELEE
Professor of Sociology, University of Missouri

Illustrated, 12mo, $2.00

This is the first book to bring together the results of the most recent work in biology, zoölogy, neurology in particular, in genetic and comparative psychology, and in anthropology, showing the significance of this work for the analysis of the fundamental factors in the determination of human behavior; namely, instinct, intelligence, feeling, and the different types of social relationships.

The book contains contributions which are vital to psychologists, anthropologists, and social scientists, and will be of great value to the general reader bringing together in convenient form the results of work which is of so much significance for the study of human behavior and human nature. To those engaged in educational work it will be of great use, and will be found valuable as a college and university text-book in certain courses in psychology and sociology.

THE MACMILLAN COMPANY
Publishers   64–66 Fifth Avenue   New York

Canada

Introduction to Zoölogy

By ROBERT W. HEGNER, Ph.D.
Assistant Professor of Zoölogy in the University of Michigan

Illustrated, Cloth, 12mo, 350 pages, $1.90

Only a few animals belonging to the more important phyla, as viewed from an evolutionary standpoint, are considered in this book. They are, however, intensively studied in an endeavor to teach the fundamental principles of zoölogy in a way that is not possible when a superficial examination of types from all the phyla is made. Morphology is not specially emphasized, but is coördinated with physiology, ecology and behavior, and serves to illustrate by a comparative study the probable course of evolution. The animals are not treated as inert objects for dissection, but as living organisms whose activities are of fundamental importance. No arguments are necessary to justify the "type course," developed with the problems of organic evolution in mind, and dealing with dynamic as well as static phenomena.

College Zoölogy

By ROBERT W. HEGNER, Ph.D.
Assistant Professor of Zoölogy in the University of Michigan

Illustrated, Cloth, 12mo, 733 pages, $2.60

This book is intended to serve as a text for beginning students in universities and colleges, or for students who have already taken a course in general biology and wish to gain a more comprehensive view of the animal kingdom. It differs from many of the college textbooks of zoölogy now on the market in several important respects : (1) the animals and their organs are not only described, but their functions are pointed out; (2) the animals described are in most cases native species; and (3) the relations of the animals to man are emphasized. Besides serving as a textbook, it is believed that this book will be of interest to the general reader, since it gives a bird's-eye view of the entire animal kingdom as we know it at the present time.

THE MACMILLAN COMPANY
Publishers   64-66 Fifth Avenue   New York

The Cell
in Development and Inheritance

By

EDMUND B. WILSON, Ph.D.

Professor of Zoölogy, Columbia University

SECOND EDITION, REVISED AND ENLARGED

Illustrated, 8vo, $3.50

Since the appearance of the first edition of this work, in 1896, the aspect of some of the most important questions with which it deals has materially changed, most notably in case of those that are focussed in the centrosome and involve the phenomena of cell-division and fertilization. This has necessitated a complete revision of the book, many sections having been entirely rewritten, while minor changes have been made on almost every page.

It has therefore been considerably enlarged, and upwards of fifty new illustrations have been added. The endeavor has, however, still been made to keep the book within moderate limits, even at some cost of comprehensiveness; and the present edition aims no more than did the first to cover the whole vast field of cellular biology. Its limitations are, as before, especially apparent in the field of botanical cytology. Here progress has been so rapid that, apart from the difficulty experienced by a zoölogist in the attempt to maintain a due sense of proportion in reviewing the subject, an adequate treatment would have required a separate volume. The author has therefore, for the most part, considered the cytology of plants in an incidental way, endeavoring only to bring the more important phenomena into relation with those more fully considered in the case of animals.

The steady and rapid expansion of the literature of the general subject renders increasingly difficult the historical form of treatment and the citation of specific authority in matters of detail. This plan has nevertheless still been followed.

THE MACMILLAN COMPANY

Publishers   64-66 Fifth Avenue   New York

An Outline of
the Theory of Organic Evolution

*With a Description of some of
the Phenomena which it explains*

By MAYNARD M. METCALF, Ph.D.

Professor of Zoölogy, Oberlin College, Oberlin, Ohio

THIRD EDITION, FUNDAMENTALLY REVISED

Cloth, 8vo, Colored Plates, $2.50

The lectures out of which this book has grown were written for the author's students at the Woman's College of Baltimore, and for others in the college not familiar with biology who had expressed a desire to attend such a course of lectures. The book is, therefore, not intended for biologists, but rather for those who would like a brief introductory outline of this important phase of biological theory.

It has been the author's endeavor to avoid technicality so far as possible, and present the subject in a way that will be intelligible to those unfamiliar with biological phenomena. The subject, however, is somewhat intricate, and cannot be presented in so simple a manner as to require no thought on the reader's part; but it is hoped that the interest of the subject will make the few hours spent in the perusal of this book a pleasure rather than a burden.

In many instances matter that might have been elaborated in the text has been treated in the pictures, which, with their appended explanations, form an essential part of the presentation of the subject. This method of treatment has been chosen both for the sake of the greater vividness thus secured and because it enables the book to be reduced to the limits desired. Many of the illustrations have been obtained from books with which the reader may wish later to become familiar.

THE MACMILLAN COMPANY

Publishers   64–66 Fifth Avenue   New York

Canada

www.ingramcontent.com/pod-product-compliance
Lightning Source LLC
Chambersburg PA
CBHW051854170526
45168CB00001B/100